# Promethean Fire

## Reflections on the Origin of Mind

Charles J. Lumsden

and

Edward O. Wilson

*Illustrations by Whitney Powell*

Harvard University Press

Cambridge, Massachusetts / London, England

10  9  8  7  6  5  4  3
Library of Congress Cataloging in Publication Data

Lumsden, Charles J., 1949–
  Promethean fire.

  Includes bibliographical references and index.
  1. Sociobiology.  2. Brain—Evolution.  3. Genetic
psychology.  4. Social evolution.  5. Cognition and
culture.  I. Wilson, Edward Osborne, 1929–
II. Title.
GN365.9.L86  1983     304.5     82-15840
ISBN 0-674-71445-8

# Preface

WHAT WAS THE origin of mind, the essence of humankind? We will suggest that a very special form of evolution, the melding of genetic change with cultural history, both created the mind and drove the growth of the brain and the human intellect forward at a rate perhaps unprecedented for any organ in the history of life.

The technical development of a theory of gene-culture coevolution, which is grounded in the facts, principles, and mathematics of biology and the social sciences, was the subject of our earlier work, *Genes, Mind, and Culture* (1981), a monograph written for scientific specialists. As we and others have pursued the nature of gene-culture coevolution from several directions, we have come better to understand its potential significance for a broad range of problems involving human behavior and social structure. We believe that an explanation of this postulated evolutionary mechanism will be of interest to a larger audience with a nonprofessional yet keen interest in human nature. At issue is the ultimate nature of man as it might eventually come to be interpreted with the aid of scientific investigation.

For the first time we also link the research on gene-culture coevolution to other, primarily anatomical studies of human evolution, and use the combined information to reconstruct the actual steps of mental evolution. We explore the implications of these and related ideas for the development of a more potent human science, which can serve as the basis for informed social action and new techniques in moral reasoning.

A note on style: we have taken this opportunity to describe the personal history of our efforts, from the hopes and frustrations that led us to conceive of the problem in the particular way we did to the steps we invented in an attempt to solve it. The narrative is woven into the explanation of the theory itself, in the belief that ideas are more interesting if given recognizable people to inhabit. We were further encouraged to adopt this method by the fact that personal accounts of theoretical research, unlike those of experimental and field research in science, are relatively rare and may have some interest beyond the purely confessional.

And a note on perspective: we are well aware of the difficulties accompanying any explanation of the origin and meaning of the mind, especially when the attempt is held to rules of scientific evidence and reasoning. So it is appropriate to stress at the outset that the scientific study of gene-culture coevolution has just begun. Furthermore, very little is yet understood of the physical basis of mind, and still less about the mind and behavior of primitive men. We have been careful to distinguish fact and true theory from mere speculation but have not hesitated to use a blend of all three to express current knowledge and—more important—to suggest additional ways by which the subject can be advanced. We are also sensitive to the concern of many scientists and others that such efforts might if successful generate sinister new forms of social control, by which is meant control of one group by another. In the final chapter we will argue that research on the biological basis of behavior increases individual freedom and reduces the opportunity for science to play a tyrannous role. Thomas H. Huxley, the biologist and natural philosopher who championed Darwin, expressed the guiding rule when he said that we must learn what is true in order to do what is right.

Finally, we wish to express our gratitude to friends and colleagues who have advised us in the course of the book's preparation: Elso S. Barghoorn, Napoleon Chagnon, Robert M. Fagen, Bert Hölldobler, Kathleen M. Horton, Peter Marler, David Pilbeam, Charles Wagley, Arthur Wang, and J. J. Yunis. The illustrations were drawn by Whitney Powell, an artist who is also a specialist in ethnographic illustration. In preparing the artwork for this book, Powell was able to make effective use of her years of experience at the Peabody Museum of Harvard University.

# Contents

# The Fourth Step
# of Evolution

THERE IS A MISSING LINK in human evolution about which few facts are known and little has been written. It is not any one of the intermediate forms connecting modern man to his primitive, apelike ancestors. About the bodies and habits of these creatures we already know a great deal. The crucial bones are being uncovered and the evolutionary trees pieced together across four million years of geological deposits. The missing link is something much more challenging—the early human mind. How did it come into existence? *Why* did it come into existence?

The mind of present-day man is the most complex process on Earth, an ultimate object of inquiry and the promised meeting ground of science and the humanities. A host of disciplines are converging on it: neurobiology, biochemistry, endocrinology, genetics, developmental biology, cognitive psychology, linguistics, computer science, cultural anthropology. The brain—the mind's machine—is being dissected down to the individual nerve cells and the molecular triggers that activate them. The events of perception and thought are being traced inward from the eyes, ears, and other sensory organs to the association areas of the cortex and back out again to the muscles and other effector organs. Using data from new and sophisticated experiments, scientists now speak cautiously of the biology of language, creativity, and other diagnostic human processes.

1

What has been missing from this enterprise in curious dispro-
portion is an organized search for the origin and evolution of the
mind. The modern synthesis of evolutionary theory, which
joined genetics to the remainder of modern biology during
roughly the half century from 1930 to 1980, was not stretched to
include psychology or any significant part of the social sciences.
To many of the wisest of contemporary scholars, the mind and
culture still seem so elusive as to defeat evolutionary theory and
perhaps even to transcend biology. This pessimism is understandable but, we believe, can no longer be justified. The mind
and culture are living phenomena like any other, sprung from
genetics, and their phylogeny can be traced. In the original sense
of the word, that history can only be marvelous.

What is the mind? We cannot yet precisely define this process
that defines us. Scientists who have made the most strenuous efforts to resolve the problem can only point to a series of key operations that distinguish mental activity from all other biological
processes in modern human beings. That much accomplished,
we can search for similar operations in apes and other animals
and try to deduce the form they took in the early human forebears. Conscious human thought entails first of all the massive
retrieval of information from long-term memory. Somehow the
information is pieced together, in substantial degree as symbols
and words, to create a map of the world as it exists outside the
brain. We screen and organize the stimuli that pour in on us in
each given instant; all that pass are refigured to represent the real
world of this instant in time. But the mind perpetually labors to
accomplish a great deal more. It recreates scenes of the past. It
invents scenarios of the future. These alternating time frames are
simulacra of the internal map of the present world, and they are
linked with it in sequences that create the sense of time. The
mind clusters images into categories and designates them by
simple symbols so as vastly to speed the mapping procedure.
Also, the mind is intentional: it summons certain images that are
emotionally desirable and toward which it is therefore prone to
move, and it concocts alternative future scenarios by which

the result can be reached from present time. It is self-aware. Part of the map consists of the physical existence of the particular brain and body that generate it. Much of the intentionality consists of the thought designed to enhance the well-being of the brain and body. At the center of the neurophysiological recall and self-assembly, a maximally intense and coherent activity comprises conscious thought. Beyond, in less focused flurries of cell activity, subliminal images and stabs of feeling form subconscious or unconscious thought—fragmentary parts of the mind that affect the flow of conscious activity but do not, for the moment at least, enter its mainstream. If and when we are able to characterize the organization of these various processes and identify their physical basis in some detail, it will be possible to define in a declarative and unambiguous manner the urgent but still elusive phenomenon of mind, as well as self and consciousness.

The evolutionary reconstruction of the mind should be pressed to the limit of this understanding. Let us start the effort, then, by going back literally to the very beginning.

OVER FOUR MILLION years ago, a small apelike animal in Africa shifted from an arboreal existence in tropical forests to a more terrestrial existence in glades and savannas. In the course of this ecological adaptation the ancestral primate evolved into an erect, bipedal form able to carry burdens in its arms and hands for long distances. From fossils three to four million years old, the animal origins of the human line can be more clearly perceived. At least one pre-man and probable forebear, *Australopithecus afarensis*, wandered the savannas and woodlands during this long interval. The adults stood no more than four feet and had a cranial capacity of about 400 cubic centimeters, the same as that of chimpanzees. Footprints of a primitive hominid—presumably *A. afarensis*—have been discovered in hardened volcanic ash at Laetoli, Tanzania. The creature that made them strode in a bipedal, flat-footed manner similar to modern man's.

By two million years ago the early hominid populations had split into at least three distinct species. Two were man-apes bearing the scientific names *Australopithecus boisei* and *Australopithecus robustus*. Both were about five feet tall and generally similar to *A. afarensis*, but heavier in form and possessed of massive jaws, a gorillalike bony crest that anchored enormous jaw muscles, and molars an inch across. These anatomical peculiarities suggest that the robust man-apes were vegetarian, able to crack seeds and shred tough vegetation as the gorilla does today. The third hominid species was a "true" man, a member of the genus *Homo* and the probable direct ancestor of modern human beings.

Imagine that we could travel back two million years to walk on the African savanna, say below the Lemagrut and Ngorongoro volcanoes in what is today Tanzania. With a little effort—perhaps walking ten miles over a period of one or two days—we time travelers would find our distant forebears. They exist as little bands of hunter-gatherers widely dispersed among the antelopes, sivatheres, saber-toothed cats, baboons, and other mammals swarming over the African plains. We belong to the species *Homo sapiens*. By virtue of the immense time span and anatomical differences, they belong to a different species, *Homo habilis*. The Latin names mean respectively "wise man" and "handy man," phrases well chosen as we shall see.

A typical adult *Homo habilis* would be instantly recognizable as human. Decidedly on the small side, just under five feet and weighing well under 100 pounds. Muzzle short, canines reduced in comparison to those of apes, incisors relatively small and vertically implanted, lower jaw trim, brow ridge bulging but not apelike. Posture erect, legs straight, chest flattened, pelvis basin-shaped to hold the downward thrusting viscera, and occipital condyles rotated forward to form a pivot on which the ovoid head is balanced. Thumbs excessively long even for a primate, hands remarkably flexible, feet narrow and long, toes shortened into stubby levers. *Homo habilis* strides erect over the grassy parkland, eyes forward, his slender arms swinging free around the waist. He holds chiseled rocks and a dead animal in clenched fists.

A closer look would create what psychologists call cognitive dissonance, a conflict of perceptions generating some amount of mental discomfort. The modern human mind strives to classify the most fundamental things of life into neat opposites: male/female, outsider/insider, sacred/profane, human/animal, and so on through the catalog of vital entities. We are loath to confuse these opposites with ambiguous examples. But dissonance in this case is inevitable, because *Homo habilis* has such an unexpectedly small and sloping cranium. His brain volume is between 600 and 800 cubic centimeters, half that of the average modern *Homo sapiens* but half again as great as that of a chimpanzee. It is possible that he can speak a very primitive language, roughly as complex as that of a modern two-year-old, enough to describe the nature and location of food and call for help. But even that amount of intellectual ability is questionable. The important point is that this earliest true human being is an evolutionary mosaic. We can describe *Homo habilis* without serious distortion as the head of an intelligent ape riding atop the body of a man.

From this vision of *Homo habilis*, painstakingly assembled by anthropologists from fragments of fossilized bone and the faint traces of ancient campsites, the ancestral populations can be judged to have reached a turning point. Although technically classifiable as men according to the diagnostic traits employed by anatomists, they had only begun to take the enormously complicated sequence of steps leading to a truly human brain and mind. For the traits we deem most important, *Homo habilis* was in the act of crossing the line separating animals—or more precisely prehumans—from true men. Only afterward did the brain grow large, evolving slowly at first and then at an accelerating rate. This development gradually transformed *Homo habilis* into the intermediate species *Homo erectus* (about 1.5 million years ago) and then *Homo erectus* into modern man. *Homo habilis* is the species in which the brain size began significantly to exceed that of all other primates. But size is only a crude indicator of the real changes that were occurring. Various parts of the brain grew at very different rates. In modern man the neocortex, seat of language and other higher cognitive functions, is 3.2 times greater

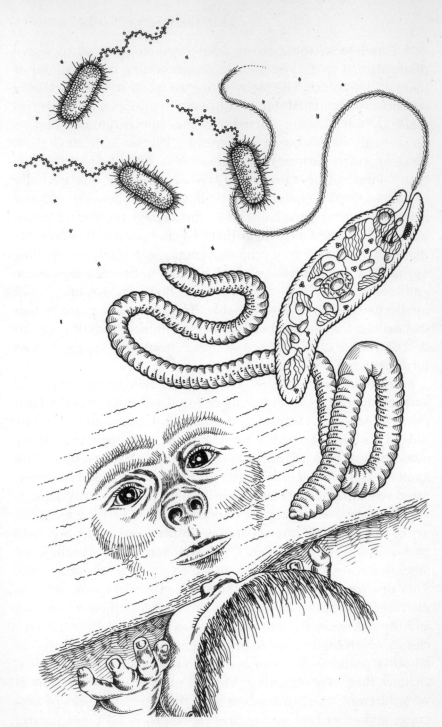

The four great steps of evolution: origin of life, first advanced cell structure, first multicellular organisms, origin of mind.

than it would be in a monkey or ape of human size. Perhaps most important, substantial but still poorly understood changes developed in the special neocortical areas concerned with speech.

The turning point at which such a fully human head and human mind began to be fashioned upon the already human body can be regarded as the latest of the four great steps in the history of life on Earth. These events, which occurred roughly one billion years apart, were first the beginning of life itself in the form of crude, replicating microorganisms; then the origin of the complex (eukaryotic) cell through the assembly of a nucleus, mitochondria, and other organelles into a tightly organized unit forming the basis of all higher life; next the evolution of large, multicellular organisms (flatworms, crustaceans), which could evolve complex organs such as eyes and brains; and finally the beginnings of the human mind.

How the last step of megaevolution was attained is a mystery of surpassing importance. Its mere contemplation brings together some of the central questions of biology and philosophy. Armed with scientific insight we can ask again with rising hopes: what is humankind, what created us, and what is our purpose in the world? The search for the origin of mind is more than just another exercise in philosophy. It goes to the heart of all of our assumptions about ethics, polity, and social purpose.

Logic and introspection give no real answers; they tell something about how the mind works but not its beginnings. Religion in turn is an enchanted hall of mirrors, a powerful device by which people are absorbed into a tribe and psychically strengthened. But in an age of scientific understanding it offers nothing concrete about man's ultimate meaning. Only a relentless search for the facts of physiology and evolution, subordinate to the self-correcting procedures of scientific analysis, seems capable of disentangling the many claims of religion. It might even lead to what the great diversity of people can agree to call the truth.

The study of human evolution consists of three enterprises. The first is the reconstruction of phylogeny, a retracing of the actual succession and branching of the hominid species. This effort depends to a large extent on the search for fossils and the skillful inference of body structure from the bony fragments recovered. A series of scientists engaged in the search have been brilliantly successful. They include Eugene Dubois, who found Java Man (*Homo erectus*); Otto Zdansky, who initiated the excavation of Peking Man (also *Homo erectus*); Raymond Dart and Robert Broom, discoverers of the australopithecine man-apes of Africa; the celebrated Leakey family (Louis, Mary, and their son, Richard), who discovered *Homo habilis* and filled in many key details of the evolution of the australopithecines; Donald Johanson, one of the discoverers and the chief chronicler of the earliest known humanoid, *Australopithecus afarensis*. Although this century-long exploration is only a beginning, the Hominidae, the scientifically defined family of man and man-apes, are already one of the best understood of any cluster of species of comparable diversity. They are as well known as horses and elephants, whose fossils are far more numerous and of better quality.

This work has identified Africa as the cradle of human evolution. During most of its recent geological history, from the Mesozoic Era to approximately 15 million years ago, Africa was cut off from Europe to the north and Asia to the east by the great Tethys Sea, a body of shallow tropical water that connected the Atlantic and Indian oceans. The present Mediterranean Sea is the final remnant of this immense waterway. Africa was an island continent similar in geographic isolation to Australia and South America. Like these land masses, it developed a distinctive mammalian fauna: elephants, hyracoids, tenrecs, giraffes, barytheres, elephant shrews, and the man-apes and earliest true men. Some of these groups were entirely indigenous to Africa. Others, including the big cats and the primates, were widespread across Europe and Asia and invaded Africa periodically, where occasional lines diversified into many species during secondary bursts of evolution. The man-apes and early men were one of the final products of a secondary radiation that occurred in the Old World primates.

Living organisms fill a continent to maximum capacity by the process of adaptive radiation. Species not only evolve through time; they also tend to split into two or more daughter species, which then come to occupy different ecological niches. After millions of years, when the radiational process reaches full maturity, a group such as the mammals contains a spectacular array of specialists. Hoofed ungulates (or in Australia, kangaroos) graze in the grassland, where they are stalked by far distant cousins who are now large predators with fangs and claws; mice or their mouselike equivalents nibble on seeds and rush to safety down burrows to escape cats or catlike creatures; moles or molelike animals burrow through the soil in search of insects; and so on through twenty or thirty major categories.

We know that during the early Miocene epoch, 20 to 18 million years ago, the common forerunners of man and the great apes lived in tropical forests. By the middle Miocene, 15 to 12 million years ago, their environment was more seasonal. Within this adaptively radiating group, the species directly ancestral to man, either *Australopithecus afarensis* or some form antecedent to it, shifted from a more arboreal, apelike existence to life spent largely on the ground. The transition was completed by no later than four million years before the present (B.P.). As part of its specialization, the ancestral form acquired a unique upright posture, bipedal locomotion, and free use of the hands. A subsequent species division occurred about 2.5 million years ago, producing the two vegetarian man-apes and *Homo habilis*, which retained a generalized diet.

This concept of species formation in phylogeny leads us to the second major theme in the study of human evolution, the reconstruction of the behavioral ecology of early men. By examining the plant and animal fossils associated with the remains of the man-apes and early *Homo*, it is possible to deduce a great deal about the climate and habitats in which the human ancestors lived. By considering the distinctive qualities in the biology and behavior of present-day human beings, especially those still living in the economically primitive hunter-gatherer societies, anthropologists have inferred some of the ways by which the human ancestors adapted to their environment. Finally, by examining ancient campsites these scientists have drawn additional conclusions concerning population density, food habits, and the material culture.

The deduction of living habits from such indirect evidence is far more difficult than the reconstruction of anatomy from bones. The hypotheses of behavioral ecology are accordingly more speculative and rapidly shifting. If any consensus can be said to exist among anthropologists, it includes the following main points. The unique upright posture and bipedalism of the hominids must be at least in part an adaptation to terrestrial life. Modern human beings normally walk at a rate of about 4.5 kilometers an hour, a leisurely but respectable pace. They do so with greater energetic efficiency than chimpanzees, who travel on their feet and the anteriorly rotated knuckles of their hands. What bipedalism accomplishes is to free the arms and hands. It is commonly argued that free hands are an adaptation for tool use. This is correct and a commonsense conclusion, but it is not the whole story. Chimpanzees also use a wide variety of simple tools, and they do so simply by standing or squatting to free their hands. Bipedalism further makes it possible to carry objects long distances, one of the most distinctive and universal activities of contemporary human beings. Did the early hominids in fact carry tools and food from one place to another, as hunter-gatherers do today? There is some evidence, from the research by the Leakeys and others, that the inhabitants of the Olduvai region in Tanzania used campsites. In this respect they would not have differed from modern hamadryas baboons, who return nightly to cliffside sleeping sites, or the savanna-dwelling anubis baboons, who travel to particular clumps of trees for nocturnal shelter. But where the baboons feed principally on seeds and other vegetable material consumed on the spot, the early hominids collected animal prey and carried them to the campsites, just as Eskimos, Australian aboriginals, and other hunter-gatherers do today. They might also have transported tubers, fruit, and similar bulky vegetable foods in the same manner.

The value of transport would have been enhanced if there were a division of labor, with some adults remaining at the base camp to protect and tend infants while others foraged. Our closest phylogenetic relatives, the monkeys and anthropoid apes, do not employ such a division of labor; modern hunter-gatherers do. Sociobiologists have pointed to the remarkable properties of human sexual behavior that appear to mesh closely with this basic ecological adaptation. Human beings are among the very few higher primates that establish long-term sexual bonds, with the male assisting the female in childrearing. The nearly total suppression of estrus, or period of heat, results in the potential accessibility of the woman throughout the menstrual cycle. It permits her not only to grant frequent sexual pleasure to her partner but also to make paternity more ambiguous and its certification a correspondingly demanding task for the male. The male baboon or chimpanzee, in order to ensure paternity, has only to note the swelling of the female's buttocks and her heightened receptivity and then to monopolize intercourse during the several days of her heat. But the human male has no such physiological guideposts and is always kept guessing. If he shares the female's favors with others, even for short periods of time, he may well lose paternity and forfeit all his investment of time and effort. Hence human males are inordinately concerned with the intricacies of courtship and pledges of reliability and good intention toward the offspring. They are enraged by signs of infidelity on the part of the female, especially when children are being conceived; cuckoldry is the leading cause of murder within hunter-gatherer bands such as the !Kung Bushman of the Kalahari. For their part, human females possess the means to judge the ardor and intentions of the males and thus lessen the chances of desertion at the early, critical period of childrearing.

Human ceremonies and forms of communication serve not only exaggerated sexuality but also the niceties of food sharing. The preparation and serving of food are key elements of social ritual in virtually every human society. Food exchange is also one of the earliest forms of social behavior in children. Two-year-olds pretend to feed their parents and greet strangers with food and other imaginary objects. These behaviors persist into nursery school, where such ritualized uses of food are routine. The early development of complex food-sharing behavior is so widespread among both primitive and advanced human cultures as to suggest that it is one of the basic tendencies of mental development. In the *slametan* feasts of Java, for example, kin and neighbors gather to celebrate some special event, marked by the choice of particular appropriate foods. Dead ancestors and supernatural beings are also invited. The social bonds of the village are reinforced, and food is redistributed in an efficient and healthful manner.

In summarizing briefly what is known about early human evolution, let us return first to the principal features of phylogeny. Through the discovery of fossils across a geological span of four million years, anthropologists have established that the earliest manlike forms, probably represented by a single species of man-apes (the most likely candidate is *Australopithecus afarensis*) first attained an erect stature and bipedal locomotion. This important step was accompanied by the acquisition of a suite of other modifications in structure and size of the teeth, conformation of the skull, shape and angle of the pelvis, and form and flexibility of the arms, hands, legs, and feet.

Behavioral ecologists have gradually assembled a theory to explain why the advance to an erect posture was taken, one that accounts for many of the most distinctive biological traits of modern man. The earliest man-apes shifted out of the tropical evergreen forests into more open, seasonal habitats, where they became committed to an exclusively terrestrial existence. They constructed base camps and became dependent on a division of labor, by which some individuals, probably the females, wandered less and devoted more time to the care of the young; others, primarily or exclusively the males, dispersed widely in the search for animal prey. Bipedalism conferred a great advantage in open-country locomotion. It also freed the arms, permitting the ancestral man-apes to use tools and to carry dead ani-

mals and other food back to the base camp. Food sharing and related forms of reciprocity automatically followed as central processes of the social life of the man-apes. So did close, long-term
sexual bonding and heightened sexuality, which were put to the
service of rearing the young. Many of the most distinctive forms
of human social behavior are the product of this tightly inter-
woven complex of adaptations.

Thus we return to *Homo habilis*, whose relatively small brain
was transported in startling configuration on top of an erect
frame and whose limited mental powers served the new advances in reciprocity and social structure. We can now examine
the third aspect of human evolution, one that extends beyond
phylogeny and behavioral ecology. That is the origin of the
unique powers of the human mind and its new partner, human
culture. To understand this phase of evolution, which carried
*Homo habilis* into *Homo erectus* and then *Homo sapiens*, is to understand the real origin of man—not just the concatenation of circumstances that made his creation possible.

Considerable attention during the past twenty years has been
rightly focused on the fossil record leading up to *Homo habilis* and
the conditions under which such a peculiar mammal could have
been assembled in the first place. But all the explorations and
speculative reconstructions have pointed solely to what biologists call a preadaptation, that is, an evolutionary change that
adapts organisms to one set of environmental conditions but in
addition and quite fortuitously positions them for a new surge in
adaptive evolution. The early australopithecine man-apes
adapted to a more terrestrial existence in open habitats by means
of a combination of particular anatomical and behavioral traits.
These qualities also preadapted their descendants for the evolutionary breakthrough that lead to the development of a very
large brain and the human mind—in other words, they set up the
main event.

In *Homo habilis* we see no more than the beginning of the major
second phase. If all hominids had been extinguished two million
years ago, man-apes and *Homo habilis* alike, their achievement to

The three enterprises in the study of evolution: the tracing of the succession of species with the aid of fossils (phylogeny), the reconstruction of the adaptation of ancient man to the environment (ecology), and the analysis of the processes that guided the evolution of the mind.

that point would have been unremarkable. Suppose that the hominids had become extinct, and eons later some new mammal group evolved higher intelligence and the ability to analyze fossils. If they were to dig up the early *Homo* remains (as our own paleontologists are now doing), they would correctly classify the hominids as just one more interesting side branch of the African mammalian radiation. That is why conventional studies of the "origin of man," while essential to the full picture, have for the most part dealt only with the conditions and circumstances that made the genesis possible. The evolutionary twists and turns culminating in *Homo habilis* comprised a remarkable odyssey indeed, but the climb from *Homo habilis* to *Homo sapiens* was by far the more momentous journey.

The creation of modern man—by which we mean the fashioning of the brain and mind—took approximately two million years, from the emergence of *Homo habilis* to the appearance of the most advanced forms of *Homo sapiens* during the past 100,000 years. The elaboration of the human brain was very rapid, perhaps the fastest advance recorded for any complex organ in the entire history of life. Even so, it merely set the stage for a still more rapid change in behavior. Almost simultaneously, culture broke past the ape level. Its growth began a slow and sometimes halting acceleration. At the time of *Homo habilis* flaked stone tools were in use, including crude forms of knives and larger choppers. By 350,000 years B.P. *Homo erectus*, the intermediate species between *H. habilis* and *H. sapiens*, employed fire and ochre pigments. Around 60,000 years B.P. Neanderthal men, racial cousins of modern man (hence the middle term in the scientific name *Homo sapiens neanderthalensis*), buried their dead with flowers, arguably the first evidence of religious belief. By 25,000 years B.P. Cro-Magnon men were producing elaborate paintings of animals and human beings on cave walls, as well as ivory figurines. They invented rituals and perhaps even myths and more elaborate forms of religion. They scratched clusters of lines on bones and rocks, possibly to keep tallies of band membership and to record the rank of leaders.

The tempo of change now quickened. Agriculture and animal husbandry were established in Asia Minor, Palestine, and the hills east of the Tigris River by 7000 B.C., and then were transmitted outward over a period of centuries to reach points as distant as Britain and China. As a result human populations vastly increased in density, congealing into villages and then into city states. That concentration and the increase in the complexity of economic transactions led to the birth of civilization, a phenomenon truly new under the sun. The first ideographic writing appeared in the Middle East by 3000 B.C. It was composed of cones, spheres, and triangles impressed into clay tablets to count stores and record transactions.

With the coming of the modern age around 1500 A.D., mankind shifted pace once again and began its final rush to the space age. Most of this advance has come through science and technology and can be fairly measured by the indexes used by historians of science. For the past three hundred years scientific knowledge has been growing at an exponential rate, essentially like a colony of rabbits, proliferating and doubling in size every few years. (An exponential rate means that the larger the size the faster the growth, and the faster the growth the more quickly a larger size is attained.) The number of surviving technical journals has a doubling time of only fifteen years. It has increased from just one, the *Philosophical Transactions of the Royal Society of London* in 1665, to about 100,000 at the present time. The population of scientists has expanded correspondingly, from a few dozen in the seventeenth century to 300,000 today in the United States alone. Some 80 to 90 percent of all the scientists in the world who have ever lived are alive today, churning out new ideas and information at a prodigious rate. Scientific knowledge and information doubles every ten years, faster than human beings are increasing, faster than people in past generations could have dreamed possible. It has been estimated, to take one of many examples, that approximately 200,000 new mathematical theorems are published each year. Over the past three centuries science has exploded from an activity involving less than one person in a million to a powerful enterprise absorbing about 3 percent of the entire productivity and manpower of the more advanced industrial countries.

Human beings seem to be on the verge of another major turning point in evolution. The storage capacity of computers is growing swiftly, with ultimate results that cannot be clearly foreseen. Technicians can now crowd more than 100,000 transistors and circuit elements onto a single silicon chip about one quarter of an inch wide. So rapid has been the advance in techniques that by 1995 it may be possible to build a computer the size of the human brain that has the same storage capacity. Satellite telecommunications have linked the world in a way that permits virtual simultaneity in the transmission of information. For the first time an entire species can be welded together into a single informational system. We can attain what Teilhard de Chardin once called the noosphere, a network of shared knowledge that envelops the earth in a manner crudely analogous to the biosphere of living organisms.

Our scientific understanding can be compared to the spreading illumination of a black forest through which the species had been traveling blindly. (The world created us but did not reveal how or why.) In the beginning the equivalent of one candle was lit, revealing a few details of the ground and the contours of nearby trees. Then visibility expanded as new sources of light were added, only a few at first and then many more at an accelerating pace. Finally, in the last instant, during our own lifetime, in less than one ten-millionth of the duration of life on Earth, this gathering power is exploding in a brilliant flash that illuminates the greater part of the interior landscape. At last we can turn and look down the long twisting paths through which the species progressed. We see our whole world and the awesome regime of life that sustains us. The time has come when the great sphinx's riddle of the meaning of man might be solved at last. But do we really want to know the answer? Our most valued myths are at hazard. Truth can seem terrible, at first.

In fact we have no choice. The crescendo in science and technology has been paralleled in other, malevolent categories of human activity: the growth of population toward its uppermost, most destructive limit in some parts of the world; the accelerating extinction of species, to such an extent that 20 percent or more of all the kinds of animals and plants might be eliminated during the next thirty years; and the swift increase of nations acquiring nuclear arms. Ethical philosophy and psychology still lag far behind the natural sciences. They must race to cope with the problems created by technology. At the very moment that man might be able to answer the question "Who am I?", the answer has become crucial to his survival.

There are between three and ten million species of organisms alive on Earth today. They represent less than 1 percent of those that have existed throughout geological history. Thus hundreds of millions of species of organisms have emerged and gone extinct during hundreds of millions of years, and of those a considerable percentage were larger animals with well-developed sense organs and at least modest brains, such as squid, giant scorpionlike animals called eurypterids, and mammals. Yet only a single lineage advanced the last step to high intelligence and complicated cultures. This circumstance, like so many of the ways of the world, is both obvious and wholly amazing. It is as if some power had reached in and plucked forth one lucky species out of a vast milling horde.

The human mind is constructed in a way that predisposes it to self-explanation by means of exalting parables. Reflective people wonder if it was the hand of God that intervened, or at least some incorporeal force beyond human comprehension. Could a divine spark have flown into the chemical machinery of living organisms and turned their evolution in a wholly new direction on just this single occasion? On the other hand, perhaps there was no outside intervention, but instead a plan was instilled into life at its origin four billion years ago or else into the very structure of physical law at the beginning of time. That would be "orthogenesis," straight-line or directed evolution. By this hypothetical means mankind was fated to emerge at the end in much the same way that a perfect eye is programed by the genes to be assembled in the last stages of fetal development. But if orthogenesis existed, it must have been driven by some force outside the known physical laws of the universe, following a master plan so elusive that it can be characterized only through a full description of the evolution that actually occurred—either that, or the very foundations of the physical sciences are faulty.

To accept explanations of deism or orthogenesis would be to abandon science and to halt the immense journey just before the meaning of the journey itself might be revealed. Some scientists and humanists, who represent a remarkably broad range of religous and nonreligious beliefs, would just as soon see the matter ended that way. For them the human spirit can never be accurately portrayed through materialistic analysis. The mind is extracorporeal—or at least cannot be related to the physiology of the brain that sustains it. Culture is an independent force growing as a layer on top of the biological mechanism, capable of being explained only by means of unique procedures and laws. This perception is the ultimate source of the troubling gap between the two cultures, between the hard sciences and the humanities. Some distinguished and careful thinkers in both fields of endeavor consider the difference to be permanent, a discontinuity grounded in epistemology and reinforced by fundamental differences in goals and interest.

Ours is a very different view. We believe that the secret of the mind's sudden emergence lies in the activation of a mechanism both obedient to physical laws and unique to the human species. Somehow the evolving species kindled a Promethean fire, a self-sustaining reaction that carried humanity beyond the previous limits of biology. This largely unknown evolutionary process we have called gene-culture coevolution: it is a complicated, fascinating interaction in which culture is generated and shaped by biological imperatives while biological traits are simultaneously altered by genetic evolution in response to cultural innovation. We believe that gene-culture coevolution, alone and unaided, has created man and that the manner in which the mechanism works can be solved by a combination of techniques from the natural and social sciences.

Our conception of gene-culture coevolution can be summarized very briefly as follows. To start, the main postulate is that certain unique and remarkable properties of the human mind result in a tight linkage between genetic evolution and cultural history. The human genes affect the way that the mind is formed—which stimuli are perceived and which missed, how information is processed, the kinds of memories most easily recalled, the emotions they are most likely to evoke, and so forth. The processes that create such effects are called the epigenetic rules. The rules are rooted in the particularities of human biology, and they influence the way culture is formed. For example, outbreeding is much more likely to occur than brother-sister incest because individuals raised closely together during the first six years of life are rarely interested in full sexual intercourse. Certain color vocabularies are more likely to be invented than others because of other, sensory rules entailing the manner in which color is perceived. Mathematical models created from the theory allow the prediction of patterns of cultural variation from a knowledge of such epigenetic rules. It is possible in principle to go from data in cognitive psychology to data in cultural anthropology and sociology, and then to work back in the reverse direction.

This translation from mind to culture is half of gene-culture coevolution. The other half is the effect that culture has on the underlying genes. Certain epigenetic rules—that is, certain ways in which the mind develops or is most likely to develop—cause individuals to adopt cultural choices that enable them to survive and reproduce more successfully. Over many generations these rules, and also the genes prescribing them, tend to increase in the population. Hence culture affects genetic evolution, just as the genes affect cultural evolution.

This particular form of the theory has been based on the research of many psychologists, geneticists, and other specialists, and it is currently being critically examined and tested. Eventually it will be either discarded or improved and extended. Meanwhile it has served to focus attention on the vital and relatively neglected relation between biological evolution and cultural evolution.

The circumstances in which we and others came to see the problem and the way in which we have tried to solve it are subjects treated in the remainder of this book. The combined effort has been a chaotic process characteristic of all scientific endeavor. It started with the discovery and synthesis of new facts, proceeded to the creation of new ideas, and was met by challenges and counter-explanations. The enterprise has advanced to the point where the new understanding can be applied as a tool of thought to the problem of the origin and evolution of the human mind.

Every such dialectic or process of scientific development must have a central theme, and in this case it was the field of sociobiology. In both history and substance, gene-culture coevolution can be understood only with reference to the influence of that new and controversial discipline.

# The Sociobiology
# Controversy

WHY HAS THERE BEEN a controversy over sociobiology? Nothing about this extraordinary episode makes sense until the subject itself is accurately characterized. Contrary to its usual popular image, sociobiology is neither a particular theory of behavior nor a politically defined doctrine on human nature. It is a scientific discipline and as such is defined as the systematic study of the biological basis of all forms of social behavior (including sexual and parental behavior) in organisms, up to and including man. General sociobiology, covering the facts and theories for all living creatures, can be usefully distinguished from human sociobiology, which addresses the topics peculiar to man. Most public interest and the bitterest disputes have centered on human applications. But the great majority of sociobiologists are only marginally concerned with this part of the discipline. They are principally zoologists, students of animal behavior who work on various social animals from colonial jellyfish to ants and chimpanzees.

Sociobiology is closely allied to ethology, which can be defined loosely as the study of whole patterns of behavior under natural conditions. Both disciplines pay close attention to the evolutionary history of species and the manner in which behavior (instinct in particular) adapts organisms to their environment.

23

But where ethology focuses on the details of individual behavior, including the activity of the nervous system and the effects of hormones, sociobiology concentrates on the most complex forms of social behavior and the organization of entire societies. Ethology consists to a substantial degree of the study of physiology and anatomy, where sociobiology is grounded in population biology: in the genetics, ecology, age structure, and other biological traits of whole aggregates of individuals.

## Evolution by Natural Selection

Most of biology is concerned with "how" questions: how cells divide, how protein is digested, how genes prescribe information. Sociobiology concentrates more on "why" questions: why cells divide in a certain way or why parents behave altruistically toward their offspring. The query "why?" can be answered only by the study of history. And the history of biological process is by definition evolution. Its creative process is natural selection, sometimes referred to as Darwinism.

As a consequence, much but not all of sociobiology consists of evolutionary explanations of such forms of social behavior as altruism, cooperation, and aggression, with special emphasis on the role of natural selection. Evolution by natural selection occurs through the following steps.

- Individuals vary in a certain trait, for example the ability to taste a particular poison, owing to differences in genes at one or more locations on the chromosomes. Let us suppose that one kind of gene (located, say, on a particular spot on chromosome number 6) confers the ability to taste the poison, while the opposing gene provides no such competence.
- The presence of the poison in food or water is the selection pressure. Individuals with taster genes detect the substance, avoid it, and survive. Those possessing nontaster genes consume the poison and die. As a result, the frequency of taster genes increases, and a larger percentage of the population has the hereditary ability to avoid the poison. This change

from one generation to the next constitutes evolution by natural selection. Individuals able to taste the poison are said to have "superior genetic fitness." The taster trait is spoken of as "adaptive."

• New kinds of genes can arise suddenly in the population by means of mutations, which are either random changes in the chemical composition of preexisting genes or rearrangements of the genes by alterations in the structure or number of the chromosomes. Suppose that in the beginning a population consists exclusively of individuals with nontaster genes. Taster genes might then originate in the population by means of mutations in one or a few individuals. If the poison occurs in food or water, natural selection would ensue, causing the taster genes to increase relative to the nontaster genes throughout the population. In a word, mutations produce the raw materials of evolution, while natural selection gives direction to evolution by determining which mutations will prevail. Incidentally, when man deliberately selects certain genes over others to produce desirable traits in plants and animals, the process is called artificial selection. When he tries to do so with himself, it is called eugenics.

Evolution sometimes proceeds by means other than natural selection. Mutations can occur at such a high frequency as to push up the percentage of mutants in the population without the aid of natural selection. Alternatively, immigrants can bring new genes into the population at a high enough rate to change the overall genetic composition of the population. These auxiliary phenomena occur and are occasionally important, but most biologists agree that they are much less potent than natural selection in directing evolution over long periods of time. In other words, natural selection is the dominant mode of evolution.

Sociobiology has extended this basic model of evolution, which is generally accepted among biologists, into the realm of social behavior. The procedure has been very successful in analyzing and explaining certain complex and previously little understood phenomena in animals. When the method is applied to human beings, the results are much less clear-cut, and they tend to provoke emotional responses. The purpose of the theory of gene-culture coevolution is to permit a deeper and more satisfactory expansion of biology into the domain of the social sciences. We will return to the relation of this theory to general sociobiology, but it would be better first to examine the classic explanations of several of the universal forms of human social behavior.

## The Selfish Genes

Individuals do not duplicate themselves during the process of reproduction. They duplicate their genes and then scatter them like seeds through the population. This strange process can be clearly visualized with the help of the following arithmetical argument. You receive one half of your genes from each parent, one fourth from each grandparent, one eighth from each great-grandparent, and so on back in a regular geometric progression. An ancestor from the late eighteenth century has bequeathed only about 1 percent of his genes to you by direct descent. Thus our individual set of genes, the hereditary material that makes each one of us biologically distinct, dissolves into smaller and smaller packets as it is traced back into history. And in a perfectly symmetrical pattern, that same set will dissolve again as it is spread among our descendants. Each child gets one half of its genes from a given parent, each grandchild receives one fourth, and so on. The only unit not ultimately divided in this manner is the gene. At each moment of the continuing present, individual human beings are of course the precious be-all and end-all, but over many generations they can be seen in a different light—as the temporary vehicles by which genes are multiplied and disseminated. From the viewpoint of evolutionary theory, and hence sociobiology, all of the traits of individuals are potential enabling devices for the expansive replication of the hereditary material that prescribes the traits. Color vision, pulse rates, insulin production, linguistic competence, the tendency to enjoy music, and tenderness toward children are creations by which the body and mind may grow strong and spread the genes through future generations. In this special sense, the genes compete in an ultimate race. Those that prescribe full color vision are matched against the various forms that cause color blindness. If color blindness declines in the population due to the handicap it imposes, so do the genes that underwrite it.

The genes come together in new combinations in each generation. Your father may have had blue eyes, the ability to roll his tongue into a tube (30 percent of the population lacks the hereditary ability to do this), AB blood type, and so on across thousands of categories of human traits controlled by particular combinations of genes. But this distinctive combination—your father—was broken into genetic fragments and reassembled into new combinations during the process of sperm production that led to your own creation. As sperm are formed, pairs of chromosomes resembling each other exchange genes and then pull apart to reduce the number of chromosomes by one half. Following insemination, a sperm cell unites with an egg to restore the original chromosome number (two of each kind), thus creating a new individual. In summary, the father's hereditary material undergoes the two basic processes of Mendelian heredity: segregation, in which the pairs of similar chromosomes are pulled apart and placed in different sperm cells, and recombination, in which matching chromosomes from the father and mother are joined to restore the original double number.

## The Meaning of Sex

The processes of sex-cell formation and fertilization are not primarily devices for reproduction. In order to replicate themselves, the genes are not required to separate from each other in a way that destroys the integrity of individual heredity. Nature has far more direct and efficient means of reproduction that preserve the entire ensemble of genes intact, such as the production of embryos from unfertilized eggs. Furthermore, sex does not merely serve the purpose of giving pleasure. The exact reverse is the case: the feeling of pleasure in the brain makes the performance of sex more likely and allows the packets of genes to be taken apart and put together again.

The primary role of sex is more subtle than straightforward reproduction: it is the creation of genetic diversity among offspring. An organism that reproduces without sex, say by hatching unfertilized eggs, can replicate itself exactly, gene by gene, without wasting time on courtship. But if all the offspring are identical, they are less likely as a group to withstand important changes in the environment. Suppose that a disease sweeping through the area kills all individuals with the mother's hereditary makeup. If the mother had reproduced in a nonsexual manner, she and all of her offspring would perish. But if she had mated with a male bearing disease-resistant genes, at least some of her offspring would survive. Also surviving would be the tendency to reproduce sexually; sex itself can be said to be favored by natural selection. Sex is slower than nonsex, but it provides a balanced array of genetic combinations to present to the world. It spreads the hereditary investment, including all the time and energy that go into reproduction, in a way that copes more consistently with harsh and constantly changing environments. Most biologists agree that adaptability, the general ability to adapt, is just as important as adaptiveness, the actual set of responses made by organisms to the environment that keeps them alive and allows them to reproduce. This long-term property is what has given sex an edge through eons of evolution and fixed it in the biology of most kinds of organisms.

We come now to the question of the differences between the sexes. When sexuality is examined in various types of organisms, from lower plants and animals to man, the fundamental distinction between males and females is seen to be based not on any set of outward anatomical traits but on the more basic character of the sex cells. Simply put, females produce large sex cells (eggs) containing yolk for the nourishment of the embryo, while males produce small sex cells (sperm) designed exclusively for the fertilization of the eggs. There is a cooperative division of labor: the female typically nourishes the embryos and sometimes goes so far as to carry them in her body or to build a nest for added protection. The male fertilizes the eggs and in a few species stays around to help care for the young. At the absolute minimum, the female and male create various combinations of their personal genes to face the unknown rigors of the future environment. Further commitment to the young is an option in evolution chosen by a minority of species, of which one happens to be *Homo sapiens*.

Because females contribute more to each fertilization, they are able to participate in fewer procreations. The average woman launches only about 400 eggs into her uterus during her lifetime, but every physiologically normal man releases millions of sperm each time he ejaculates. As a result, one male has the capacity to participate in a vastly greater number of fertilizations than a female.

Evolutionary theory predicts that several important consequences will follow from this elementary difference between the sexes. Perhaps the most basic is that males as a rule have a good deal more to gain by competing for mates. A Don Juan can theoretically become a father every night. But if he succeeds in achieving many fertilizations, a corresponding number of males must fail to become fathers. Throughout the animal kingdom, and in most human societies as well, males in fact compete aggressively with each other for territory, status, and above all access to females. There is a strong selection pressure to acquire, in Darwin's words, both "the power to charm the ladies" and "the power to conquer other males in battle." This is sexual selection, a special form of natural selection. As a result males are typically more reckless and ostentatious than females, and extremely intolerant of being cheated. Yet in many species, including the human, they are equally capable of forming close and affectionate bonds with their mates and offspring, as well as friendly alliances with other males.

In most higher animal species, females also participate in sexual selection, but in a qualitatively different manner. They can be inseminated in a given reproductive cycle by only one male, and they usually have the greater burden of looking after the young. Therefore it is to their advantage to be more discriminating: flirtatious, to attract many suitors, but also hesitant, socially skilled, and perceptive in order to mate with the best of the males. By "best" in this case is meant the most competent in dealing with other males and, among those species in which males help to rear the young, the male most likely to devote himself cheerfully to that task.

In view of these circumstances it is not surprising that most animal species are polygynous—males prefer to mate with more than one female. And relatively few are polyandrous, with females forming a bond with more than one male.

## Altruism

In its narrowest conception, evolution by natural selection—Darwinism—seems to imply survival of the fittest, the triumph of some individuals over others and the perpetuation of their genes in the next generation. But surrounding this unappealing image is a soft glow of altruistic behavior. We know that parents are willing to sacrifice a great deal on behalf of their children, even their own lives. Such behavior still conforms to Darwinism in the strict sense, because so long as children are preserved the parent's genes are passed on. If the self-sacrificing behavior of the parent is prescribed by some of the favored genes, then that particular form of altruism will be spread through the population.

Altruism of a broader kind can evolve by essentially the same mechanism. Note that brothers and sisters are just as closely related to each other as parents and offspring. Put another way, any two siblings share half their genes through common descent—the same fraction as that shared by a given parent and child. If a man makes a sacrifice on behalf of his brother or sister, and as a result of his act the sibling has more children, the man's altruistic act will cause an increase in the number of genes that are identical to his own. This form of natural selection is called kin selection, and it can cause a spread of altruistic behavior toward close kin other than direct offspring. Suppose that an individual either dies or gives up any chance to have children in order to perform an altruistic service. The biological result is the same for both cases—no direct transmission of genes to the next generation. Suppose also that variations in the propensity for altruistic behavior result from corresponding changes in certain of the genes. The question is: how much good must the beneficent act do the relatives in order for such genes to spread? The answer is that the number of offspring must be increased by a factor that is the reciprocal of the fraction of genes that are identical through common descent. The relation is simple. A brother has half of his genes identical to those of the altruist. So the sacrifice of the altruist must result in at least a doubling of the number of the brother's offspring for the altruism-causing genes to be spread (the reciprocal of ½ is 2). A first cousin has one eighth of his genes identical; if the altruism is directed at first cousins alone, their offspring must be increased at least eightfold for the altruism genes to spread. And so on.

We can see why hard-core altruism, given freely and without thought of reciprocation, is usually confined to the closest relatives, in both animals and human beings. Outside the circle of first cousins, only a spectacular increase in the success of the benefited relative is sufficient to compensate for a unilateral sacrifice. In human beings and a few of the most intelligent monkeys and apes, the circle of altruism is broadened by reciprocal altruism. In this soft-core form of giving, the altruistic act is performed with the expectation that the beneficiary will repay in kind at some future date.

Sociobiology is often said to take the goodness out of altruism. This characterization, which refers to the hard-core version of the behavior, means that while an individual may give up an advantage, perhaps even his life, the genes that prescribe such behavior give up nothing; on the contrary, they gain in number and influence. This difference is not really a contradiction. It is simply natural selection working simultaneously at two levels, that of the gene and that of the individual organism. The individual can behave in a moral and unselfish manner toward others, but his conduct results in an even greater proliferation of his genes than if he acted with consistent selfishness.

## Aggression

Aggression consists of a diversity of behaviors united only by the common feature of harmful action or the threat of harm to others. Some animals display several different stereotyped forms of aggression. When facing most larger enemies, a rattlesnake coils, draws its neck and head into an S-shaped striking posture near the center of the coil, and vibrates its rattle. If the adversary is a kingsnake, a particularly dangerous predator on other kinds of snakes, the rattlesnake pulls its head beneath the coils and slaps at the intruder with an extended loop of its body. When pursuing prey of its own, such as a mouse, it neither coils nor rattles but strikes silently from any convenient position. Finally, two male rattlesnakes fighting over access to females engage in neck wrestling. Although one could kill the other with a single venomous strike, they do not. The two snakes settle the dispute with a ritualized form of fighting that usually gives the advantage to the larger and stronger contender.

The rattlesnake illustrates several important features in the evolution of aggression. First, a great deal of the behavior consists of threat and ritual. Animals expend large amounts of time and energy persuading their adversaries that they are powerful and dangerous. When fighting does occur between members of the same species, it is most commonly in the form of ritualized combat. In such exchanges the contender who is thrown to the ground, pushed out of the arena, or otherwise bested usually accepts his defeat. He leaves the field uninjured and tries his luck elsewhere. Sometimes the exchange is limited to purely visual displays. In the great majority of cases the contest is won by the individual who is being pressed on his own territory by an invader. When no such initial advantage exists, as when the contenders meet on neutral ground, it is usually the larger individual who prevails. Only when an individual is trapped and physically endangered is he likely to resort to escalated fighting, an all-out attempt to destroy the opponent.

Rattlesnakes are also typical of most kinds of animals in the opportunism displayed by their evolution. Each of the four forms of aggression they employ is clearly shaped to serve a particular function. This circumstance leads to an interesting question in evolutionary theory. If no relevant function existed in a given species, would the absence of a need result in the absence of aggression? The answer is yes, so far as we can tell. Species of animals are known that appear to lack all forms of such behavior. Aggression, apart from the seizing of prey and defense against predators, is a device to deal with competition for limited resources. When competition is lacking, the capacity for aggressive behavior is absent. Some plant-feeding insect species are kept at moderate population levels by predators, disease, and emigration. They rarely become abundant enough to run out of food, shelter, or places to lay their eggs. Such species do not "rely" on competition to regulate their numbers, and aggressive behavior has not been added to their repertory. Another way of putting the matter is that there is no generalized instinct of aggression that permeates animal behavior, as implied in the earlier writings of Sigmund Freud, Robert Ardrey, and Konrad Lorenz. Aggressive behavior is opportunistic, tending to evolve into certain forms that appear and are shaped genetically according to the particular needs of the species.

Ecologists and sociobiologists have made substantial progress in defining the circumstances under which the varieties of animal aggression evolve. Territoriality, the defense of a space by means of threats of fighting, is most likely to appear when some part of the living space of the animal contains a resource that is both in short supply and predictable in occurrence. Thus if a location contains a regular supply of food or a superior roosting site, and either of these resources is scarce enough to limit population growth, animals are likely to try to preempt that part of the space. But in addition this territory must be defensible in an economic sense; the risk to life and the energy spent to maintain it must be outweighed by the benefits gained through owning exclusive rights to it. In fact, animals tend to defend areas that are just large enough to provide them with their minimal needs on a year-round basis.

## Social Behavior

When the full sweep of social evolution is examined from bacteria to man, the following paradox emerges. As we ascend the evolutionary scale arranged on the basis of anatomy, physiology, and brain size, we descend in the quality of the traits intuitively associated with sociality, namely cooperation, altruism, division of labor, and integration.

For convenience we can recognize four pinnacles across this array, represented by the colonial invertebrates (corals, sponges, colonial jellyfish); the social insects (ants, bees, wasps, termites); monkeys, apes, and other social mammals exclusive of man; and man himself. The members of coral colonies and similar tightly clustered groups of invertebrates originate from a single fertilized egg and multiply by the simple fission and budding off of entire new organisms. As a result the individuals living together are identical in hereditary composition. So kin selection can easily overcome strictly individual selection: looking after a neighbor is just the same as looking after oneself so far as replicating genes are concerned. This circumstance clears the way for the evolution of extreme forms of social organization. The members of the advanced colonies are fitted together into a superorganism almost indistinguishable from a single, well-knit organism. Thus the Portuguese man-of-war resembles a large jellyfish but is actually made up of a cluster of many very specialized individuals. One at the top forms the bladderlike float, others farther down are bell-shaped structures that propel the colony through the water, while still others at the bottom are shaped into tentacles and odd receptacles for the receipt and digestion of food. Only a few individuals contribute sex cells for the production of new colonies.

The same considerations hold for the social insects, but to a lesser degree. Colonies of bees, wasps, and ants comprise mostly sisters. Because of peculiarities in the way in which sex is determined in these hymenopterous insects, sisters are more closely related to one another than are mothers and daughters. On the average, three quarters of their genes are identical by common descent instead of the usual one half. It is also true, probably as a partial result of the genetic similarity, that most of the females in a colony are sterile workers, who are often further modified into specialized castes such as nurses and soldiers. The insect society is far less well integrated than that of the Portuguese man-of-war and other colonial invertebrates, but it is still much more so than mammal and human societies. Its members live in a state of impersonal intimacy. The individual worker is almost always in close touch with other colony members, and virtually everything it does is directed to the welfare of the colony. Hence it can be truthfully said that one ant is no ant. On the other hand, the workers do not recognize each other as individuals. Their entire behavior is conducted with reference to the colony population as a whole, or at the very most to the queen and various castes, and not to other individual workers.

The mammalian societies represent a radical departure. The members of a monkey troop or an elephant herd are less intimately associated than are the colonial invertebrates and insects. They also recognize and treat each other as individuals. They create dominance hierarchies, pair off to mate, and form into kin groups. Whereas the worker ant always acts for the good of the colony, thus contributing to the reproductive power of the queen, the individual mammal only marginally serves the welfare of its group. In fact the opposite relation holds: social life is exploited for the improvement of individual survival and reproduction. In comparison with ant workers, the members of a mammal society are extremely self-centered, strife-ridden, and preoccupied with sexual roles.

Then (and here we encounter a real paradox) the downward

trend leading from the social invertebrates to the mammals is partly reversed in *Homo sapiens.* Human beings are unique in their possession of a fully symbolic language, an enlarged memory, and long-term contracts upon which elaborate forms of reciprocity can be based. They have attained relatively high levels of cooperation, altruism, division of labor, and social integration. All this has been accomplished without surrendering the ancient mammalian heritage of personal identity and welfare. But how? And where do the mind and culture fit into the picture? We have come at last to the root of the sociobiology controversy—or, more accurately, the sociobiology controversies, for there were two that flared up in rapid succession in the late 1970s.

## The First Sociobiology Controversy

Academic disputes are more vicious than arguments in ordinary life because the status of academic professionals depends not on money, physical achievement, or even formal rank, but on the written tokens of esteem or contempt bestowed by their peers. Academic arguments escalate into open warfare if the disputed issues embrace the meaning of life, the foundation of ethics, or, above all, political beliefs and the security of academic disciplines. Then onlookers are treated to the spectacle of what in another context has been called the fury of the theologians.

So it was with the sociobiology debate, which began in the 1970s with the writings of a group of scientists from biology and the social sciences, including Richard D. Alexander, David P. Barash, Jerram L. Brown, John Crook, Richard Dawkins, Pierre van den Berghe, and Edward O. Wilson. These efforts were designed as an introduction to a new field of study. For the first time they drew together many previously isolated threads of research and defined the discipline of sociobiology within the theoretical framework of modern biology. Sociobiology did not appear out of thin air. It followed preliminary and more philosophically oriented works on the role of biology in human behavior by Konrad Lorenz, Donald T. Campbell, Irenäus Eibl-Eibesfeldt, Desmond Morris, Robin Fox, Lionel Tiger, and others. It was based to a large extent on the achievements of hundreds of investigators, from Jane Goodall's field study of chimpanzees and Theodore Schneirla's discovery of the secrets of army ant life to theories on the evolution of altruism by William D. Hamilton, John Maynard Smith, and Robert L. Trivers.

Sociobiology has grown swiftly during the past ten years, alongside ethology and behavioral ecology. It has become the subject of five new technical journals and a spreading shelf of textbooks, research symposia, and monographs. The mainstream of animal sociobiology is a rich mixture of fact and theory about the intimate lives of fungus-growing termites, fence lizards, anubis baboons, and the tens of thousands of other species of social creatures that have attracted the attention of zoologists around the world. Very few of the findings of this field are the subject of dispute. And when the principles derived from the basic studies are extended to human beings, there is no special reason to expect that human social behavior is completely genetically determined. It could be entirely free of heredity or, more likely, something in between. The question can be addressed with equanimity through the interlocking techniques of genetics, psychology, and sociobiology.

In spite of the inherent neutrality of the discipline, academic partisans were soon locked in the win-or-die struggle usually associated with the question of heredity versus environment. The reason is that most sociobiologists did take a position on the key question. They suggested that a great deal of human social behavior is affected by heredity and hence can be explained more readily by biology than by the usual formulations of the social sciences. This proposition, although comprising only a modest part of the overall subject, received far more popular attention than the remainder of the discipline. The sociobiology debate was triggered by both the idea of genetic control (which seemed to negate the existence of free will) and the means by which the enterprise was legitimated: a more hereditarian view of human behavior had been invested with the apparent authority of a full scientific field, and the social sciences were opened to the intrusion of biology. Similar opinions had been prominently expressed in earlier books, including especially the best-selling *On Aggression* by Konrad Lorenz, *The Territorial Imperative* by Robert Ardrey, and *The Naked Ape* by Desmond Morris. But these were works designed for a broad audience, devoted to only a few topics in human behavior, and they appeared to marshal evidence to support a single idea. To critics predisposed against a genetics of human nature, they were dismissible. The debates that swirled around them foreshadowed the sociobiology controversy but died down with far less impact than the central ideas deserved. Sociobiology, in contrast, was a scientific discipline, into which human behavior had been embedded as the idiosyncratic evolutionary adaptation of one among many primate species. It was not so easily dismissible.

The principal technical novelty of sociobiology is its treatment of societies as populations, subject to the laws of organized systems above the level of isolated individuals. Each society—each population—possesses a particular size, overall genetic structure, ratio of males to females, means of communication among the members, and other higher-order features. These traits can be observed as they change through time, just as the chemistry, anatomy, and behavior of organisms are analyzed in more conventional biology. We speak of such studies as population biology, the analysis of life at the level of breeding groups of organisms. Sociobiology is to a considerable extent the special branch of population biology that deals with social animals and human beings, although it also draws heavily on other branches of biology. Above all, it is entirely orthodox with respect to the natural sciences. Its practitioners remain true to the generally accepted facts and principles of biology and attempt to draw on the full arsenal of that larger science.

On first inspection it might appear logical to apply the principles of general sociobiology to human beings. After all, our knowledge of human heredity and physiology, the very foundation of modern medicine, was built to a large extent on principles derived from the study of other organisms, including such lowly forms as fruit flies and colon bacteria. Might animal-based biology also enrich the foundations of the social sciences? This prospect seemed to appeal to the academic community, and sociobiology promised to fulfill a widely perceived intellectual need.

Then, unexpectedly, a group from the American radical left began a major campaign to change the earlier opinion. They viewed the effort to extend sociobiology to human beings less as scientific research than as the stirring of a dangerous political monster. During the summer of 1975, fifteen scientists, teachers, and students from the Boston area formed the Sociobiology Study Group of Science for the People. Most prominent among them were four colleagues of one of us (Wilson) on the Harvard faculty: Jonathan Beckwith, Stephen Jay Gould, Ruth Hubbard, and Richard Lewontin. They were interested, as Hubbard said, in "breaking down the screen of approval" springing up around the new field, particularly Wilson's recently published *Sociobiology: The New Synthesis*. After meeting quietly for several months, the Science for the People committee published a letter in the November 13, 1975, issue of *The New York Review of Books*. There they said that human applications of sociobiology are to be condemned because of the political dangers such thinking poses. All attempts to establish a biological basis of social behavior will lead to Social Darwinism and the attendant evils favored by capitalist-imperialist regimes. Even the presentation of hypotheses of this nature

> consistently tend to provide a genetic justification of the *status quo* and of existing privileges for certain groups according to class, race, or sex. Historically, powerful countries or ruling groups within them have drawn support for the maintenance or extension of their power from these products of the scientific community. For example, John D. Rockefeller, Sr. said "The growth of a large business is merely a survival of the fittest . . . It is merely the working out of a law of nature and a law of God." These theories provided an important basis for the enactment of sterilization laws and restrictive immigration laws by the United States between 1910 and 1930 and also for the eugenics policies which led to the establishment of gas chambers in Nazi Germany.

To link opponents with Rockefeller and Hitler is to call for their exile from the dovecotes of academia. The purpose is not so much to answer the arguments of the authors under attack as to destroy their credibility. This is particularly true at Harvard University, where a professor accused of fascist sympathies is in roughly the same position as an atheist in a Benedictine monastery.

The Science for the People representatives were well-meaning. They opposed human sociobiology because its very pursuit implied to them a belief in genetic determination of social behavior, and "genetic determinism" offers support for racism, sexism, and the status quo. In their view, any demonstration that differences in intelligence and temperament can be influenced by heredity lays the groundwork for treating people differently according to race or sex. To suggest that the proneness toward aggression and dominance is partly inherited is to validate these destructive human traits. The Science for the People group spoke with conviction when they claimed that *any* admission of genetic determinism can be used by the ruling classes to justify the existing social order, whether it is true or not. If the idea of genetic determinism is allowed to flourish, they said, legitimate struggles for social justice will be suppressed with the blessing of science. To oppose this kind of pernicious research and speculation, while redirecting science to the greater benefit of the poor and the oppressed, is properly designated Science for the People. Armed with this humanitarian formula, the critics attempted to take the moral high ground in the opening round of the sociobiology debate.

The flaw in their argument, pointed out by many subsequent writers during the debate, is the assumption that scientific discovery should be judged on its possible political consequences rather than on whether it is true or false. That mode of reasoning led earlier to pseudo-genetics in Nazi Germany and Lysenkoism in the Soviet Union. The Science for the People committee were on equally shaky ground in their political rationale. If societies regard racism or any other kind of injustice as undesirable, whatever the causation, its ravages can be mitigated. A knowledge of the hereditary tendency and biological mechanisms then becomes a valuable part of political reform rather than an obstruction to it, in the same way that information about the hereditary basis of sickle-cell anemia and hemophilia is essential to the diagnosis and treatment of these disorders. Opponents of Science for the People argued that to suppress scientific research on the grounds of how it *might* be misused is indefensible. They reasoned that the alleviation of human suffering will be favored on a global basis, not impeded, by a deeper investigation of the genetic processes of human nature. This procedure will steadily replace rumor and folklore with testable knowledge. Nothing except ignorance and fear is to be gained by the dogmatic denial of the existence of hereditary components and attempts to discredit the scientists who study them. Information humanely acquired and widely shared, related to human needs but kept free of political censorship, is the real science for the people.

The social aims of Science for the People, stressing justice and equality, were admirable, but the logic they used undercut their own political cause. By emphasizing the absence of hereditary influence on human behavior as a virtual precondition for social justice, they made the very idea of justice hostage to further scientific research. For if further research were to produce evidence contrary to their hopes, and genes were shown to play a role in human behavior, Science for the People would not merely lose the technical dispute. They would also find it extremely difficult to respond to those members of the extreme right who genuinely favor discrimination and the status quo—for reasons having nothing to do with respect for scientific evidence.

And that is essentially what happened. In recent years evidence has steadily mounted to implicate genes in virtually every category of human behavior chosen for close study. In order to test for hereditary influences in differences among people, geneticists have made detailed studies of twins and other siblings raised together and apart. They have supplemented this work with complex pedigree analysis and the tracing of mental devel-

opment of children reared in different environments. Their research has produced substantial evidence of hereditary variation
in color vision, hearing acuity, odor and taste discrimination,
number ability, word fluency, spatial ability, memory, timing of
language acquisition, spelling, sentence construction, perceptual
skill, psychomotor skill, extroversion/introversion, proneness to
homosexuality, age of first sexual activity, timing of major stages
in intellectual development, proneness to alcoholism, some phobias, and certain forms of neurosis and psychosis, including
manic-depressive behavior and schizophrenia.

Although human genetics is still in its infancy, researchers
have already identified the exact sources of some of the more extreme forms of behavioral alteration. Single genes have been
found, for example, that induce a compulsive tendency to pull
and tear at the body. Others lower the capacity to perform on a
few standard cognitive tests used in psychology but not the remainder, or affect sensitivity to the odors of musks and floral
scents. The mapping of the human chromosomes has begun in
earnest, and it is certain to have a profound effect on both medicine and our perception of human nature.

The basic weakness in the Science for the People argument
was unfortunately noted by the extreme political right, including
the National Front of England and the New Right of France.
These groups, who favored various shades of authoritarian government and racist practice, accepted the Science for the People
arguments with enthusiasm. They proclaimed, in effect, that if
the critiques were wrong, then the political ideals endorsed by
Science for the People must also be wrong. They embraced sociobiology—not the real science of sociobiology but rather the
caricature drawn by Science for the People and a few of the more
sensationalist popular writers. Although it was far from what
they intended, Gould, Lewontin, and other members of the Har-
vard-based group had turned political ideology into a contest
that seemingly could be settled by factual studies in human genetics.

Fortunately, the debate went in a different direction. Most

people in the United States and Europe do not care very much for ideologies of either the extreme left or right; they also pay very little attention to academic squabbles. Among those who do care, it was generally perceived that Science for the People was something more than a benevolent study group devoted to the humane uses of science—it was also an action group that promoted Marxism-Leninism, in a manner specialized to subordinate science to the service of that ideology. Its official journal, *Science for the People*, consistently praised communist regimes while excoriating the capitalist and "imperialist" states of the West, and most particularly the United States. The writers for *Science for the People* called for the exposure and harassment of scientists, scientific projects, and institutions that they classified as supportive of the prevailing order. A sample of titles from their articles during 1975–1979 conveys the flavor: "Social Science Research: A Tool of Counterinsurgency," "Sociobiology: Tool for Social Oppression," and "U.S. Medical Research: For the Power Not the People." Social criticism is a good thing, but the writers had a larger ideological purpose in mind and were not prepared to tolerate challenges to their own beliefs. Nearly all of the villains identified by them over the years live in capitalist countries; few if any represent the "new, humane" socialist countries that the activists sought to emulate. When a project in human genetics at the Harvard Medical School was closed down because of a publicity campaign by Science for the People and the resulting public harassment of its directors, Jonathan Beckwith was triumphant. Proclaiming an important victory (over the Medical School faculty, which overwhelmingly supported the project), he called for a widening of the struggle against all "ruling class academic institutions."

Given the expression of such sentiments, it was not unreasonable for the academic community at large to suppose that Science for the People were opposed to human sociobiology for reasons other than a pure concern for humanity and scientific truth. Their campaign had the look, as Tom Bethell once summarized the matter, of burning Darwin to save Marx. So by 1978 what we knew at Harvard as the political wars of sociobiology had largely come to an end.


## The Second Sociobiology Controversy

But even as the clamor died around Harvard and other campuses, a second and more substantial controversy fast emerged in full strength. Scholars who actually study human behavior, and in a manner not noticeably encumbered by outside political agendas, had discovered basic flaws in the sociobiological program. To be sure, opinion was sharply divided on the fundamental issues. Many social scientists and philosophers, including Mary Midgley, Alexander Rosenberg, Michael Ruse, Peter Singer, Donald Symons, and Pierre van den Berghe, considered the sociobiological approach to be basically sound and promising, if still largely untested. But others, including Kenneth Bock, Clifford Geertz, Stuart Hampshire, Marvin Harris, Edmund Leach, and Marshall Sahlins, argued that sociobiology cannot have much relevance for the social sciences and humanities. They held that what is unique, richly structured, and most interesting in human existence is a product of the conscious mind permanently beyond the reach of biological investigation. The natural sciences can never be joined with the social sciences and humanities because the subject matter and whole intent are fundamentally different. "The central incoherence in the idea of sociobiology," Hampshire wrote, "arises at the juncture of explanation which, serving different purposes, cannot be welded into a continuous whole . . . It is important that one should not see this irreparable break as a division in reality, but rather as a division between two divergent sets of human interests, both irreplaceable interests."

These nonpolitical critics conceded that sociobiology works well for animals. They allowed that broad central tendencies in human social behavior can be predicted to a limited extent by the zoologically based theory. For example, the treatment of close relatives is in loose accord with the principles of kin selection that have been carefully tested in the social insects. And territorial behavior is displayed by hunter-gatherers under approximately the conditions predicted by theoretical ecology. Incest avoidance is as it should be, according to the theorems of population geneticists. But this way of describing human life, the critics insisted, remains grossly inadequate. Human beings are not automata that perform simply according to the instructions of their genes. They have minds and free will. They can perceive and reflect upon the consequences of their actions. This high level of human mental activity creates culture, which has achieved a life of its own beyond the ordinary limits of biology. The principal habitat of the human mind is the very culture that it creates. Consequently, individual cultures diverge in their evolution and they vary enormously from one society to the next in ways that cannot be explained by traditional reductionistic biological analysis. The questions of importance in the social sciences—of mind, self, culture, and history—are beyond the reach of sociobiology as that subject was originally formulated.

These criticisms of human sociobiology, also forcefully argued by Science for the People, were largely correct. They came to be fully appreciated even by those most optimistic about the prospects of the new discipline. It is fair to say that by the end of 1978, when the arguments on both sides had been registered in technical journals and textbooks, the sociobiology controversy was in a stalemate.

## An Attempt at Resolution

Charles Lumsden's arrival at Harvard created an opportunity to break the impasse. Lumsden had recently finished his doctoral training at the University of Toronto and joined forces with Wilson for the first time in January 1979. A theoretical physicist trained in both the hard sciences and the human life sciences, Lumsden's forte was physical theory in biology. He explored ideas and mathematical procedures for relating the order and complexity of biological systems, the human brain in particular, to underlying principles. Lumsden came to Harvard to work on the regularities of the highest level of organic integration, that linking organism with society, which is the subject matter of sociobiology. The social insects, with their precise clockwork of communication and division of labor, seemed to him a perfect place to begin.

At first it appeared that ant societies would occupy our attention completely. From a theoretician's viewpoint they were ideal. Moreover, Wilson had recently finished *On Human Nature*, an extended essay that explained sociobiology to a wider audience and explored its philosophical implications. He was battle-fatigued from the political controversy. He had also devoted three years to research on caste systems in ants, termites, and other social insects. This work, which was conducted with George F. Oster of the University of California at Berkeley, culminated in the monograph *Caste and Ecology in the Social Insects*. The book opened a new array of fascinating problems for theoretical and experimental study.

So we were both primed for research on social insects. Our intention was to collaborate in a study of ant societies, with an emphasis on the transition from individual to group behavior among worker ants. But that didn't happen. As we sat in Wilson's office and tried to talk about insects, the conversation kept drifting back to human sociobiology and the stalemated controversy. We toyed with the elusive problems of mind, consciousness, free will, and cultural diversity. We questioned whether these phenomena are insuperable barriers to sociobiology. Does a no-man's-land really exist across which biology can never pass to the social sciences and humanities? It seemed to us very unlikely that nature is divided into such independent domains of reality. Surely to incorporate mind and culture into evolutionary theory stands as one of the great intellectual challenges.

An additional incentive for such a venture exists in the very foundations of the physical sciencies. During the early twentieth century, physicists had inherited from philosophers and psychologists a completely unexpected phenomenon in the revolutionary quantum theory of nature: the human mind. In this re-markable body of physical laws, now tested and unrefuted after fifty years of stringent experimentation with atoms and sub-atomic particles, the human mind acquired an intrinsic role. In complete contrast to preceding theories of the universe, the ob-server and his world were no longer separable, even in principle. The human mind did not stand apart from physical reality. When sifted for meaning, the mathematics of the quantum the-ory seemed to say that, through the very acts of knowing and decision, the conscious mind influences the course of events in the external world. Not everyone agreed with this possibility, and quantum wars had raged in physics at the same time the gene wars were sweeping through behavioral biology. With so little known for certain about the human mind, the mystery seemed complete. We realized the implications for this impasse that might follow from a successful attempt to bring human mind and culture within the framework of the natural sciences.

Above all, we recognized that the entrée to the whole compli-cated subject lies in the formulation of the correct *problem*. We agreed that the problem must be the exact nature of the relation between genetic evolution and cultural evolution. The two pro-cesses are somehow linked through the mind. If the steps leading from genes to mind and then to culture could somehow be char-acterized precisely, the evolution of man and the origin of mind would be seen in a new light.

Very little research was being devoted in the late 1970s to the relation between genetic and cultural evolution. The reasons are complex. There was, for example, the perceived danger of reviv-ing the old images of Social Darwinism in the minds of liberal and radical intellectuals. The sociobiology controversy was an example of the fire storm awaiting those who try to traverse the zone between biology and the social sciences. Another cause of anxiety is the traditional autonomy of the study of human beings. The founders of the modern disciplines of the social and behavioral sciences, particularly Émile Durkheim in sociology, Franz Boas in anthropology, Sigmund Freud in psychoanalysis, and J. B. Watson in behavioristic psychology, had gone to great lengths to protect their infant subjects from the biological imperium. With a few outstanding exceptions, their followers continued this tradition into the 1970s. It was widely regarded as sinful to "biologize" the social sciences; the territorial boundaries were still being defended with patriotic devotion.

A final inhibiting factor is the sheer difficulty of creating an accurate portrayal of genetic and cultural interaction. It is easy enough to write fluent essays about the prospects of such research and to warmly recite the intellectual history of failed attempts. Such is the honored tradition of social theory written as literary criticism, which is to say most social theory. It is something else to create a chain of scientific models with results that can be tested by observation. Every thoughtful person looks upon the face of human nature with the confidence born of untested intuition, just as he gazes up at the moon and once in a while imagines going there. To a large extent, social theorists and intellectuals who had claimed human nature for their domain had been taking imaginary trips to the moon. They had repeatedly postponed the implications of the plain but awesome fact that the brain is a machine created by genetic evolution.

A few other biologists perceived the importance of the subject roughly as we did. Among the most active during the 1970s were William Durham at the University of Michigan; Robert Boyd and Peter Richerson at the University of California, Davis; and Luigi Cavalli-Sforza and Marcus Feldman at Stanford. Each had produced elegant new insights and mathematical techniques touching on the interaction of genes and culture. Others, including Donald T. Campbell, Konrad Lorenz, and Martin E. P. Seligman, had written convincingly on the evolution of genetic bias in learning and human perception. But none of our colleagues had broken the key problem of the role of mind in evolution or even expressed it in soluble form. At the time we began working, psychology and the brain sciences still lay outside formal evolutionary theory, and the linkage between genes, mind, and culture was by and large untracked territory. The central problem of sociobiology was no longer the evolution of altruism, as it was in the early days of the discipline; that problem had been largely solved by powerful theoretical and experimental studies in the 1970s. The central problem was now the relation between genetic evolution and cultural evolution.

We recognized that a major shift was needed in the way human evolution is viewed, with mental development and cultural diversity taking center stage. The theory we wished to build would contain a system of linked abstract processes expressed as far as possible in the form of explicit mathematical structures that translate the processes back to the real world of sensory experience. The value of the theory would depend ultimately on two features: its internal consistency as it plays over the entire range of human existence, and the truthfulness of its hypotheses and models in dealing with the main features of culture thus encountered. The concurrence of biologists and social scientists could be honorably won only if the theory provided a novel way of thinking about human evolution that accounts for some of the key traits of mind and culture in a manner superior to competing explanations. A principal goal would be to fashion models out of the theory that correctly account for the pattern of cultural diversity. The models could eventually be tested by the observed idiosyncrasies of mental development and the properties of genetic and cultural change in various social settings. The more surprising the confirmed results, the more they contradicted intuition, the better.

Now we should confess the romantic self-image of the theoretical scientist: to travel from the life of immediate sensory experience through layers of mental abstraction sufficiently deep for the purpose of creating a coherent whole within the mind, and then to carry this new internal world back out to understand sensory experience. And this is the romantic self-image of the experimental scientist: to employ such abstractions in the gathering of new knowledge and—should the theories prove unequal to the task—to destroy them remorselessly. Concepts formed by the scientific imagination, as the physicist Gerald Holton has said, "are not some distillation of the experiences which any-body, using the kind of logical reasoning one supposedly learns in school, should sooner or later be able to trace." On the con-trary, "the concepts themselves are freely formed, subject only to the *a posteriori* usefulness of the whole structure when confronted with experience."

Our goal was to produce a theory that could survive and draw growing strength from new generations of models. At the heart of the enterprise was an unknown mechanism that we believed has to exist. The mechanism must reveal itself in the patterns and diversity of human life. We would try to capture it in whatever modifications of the evolutionary process are required to fit the system to human beings. The nature of the mechanism was the problem to be solved.

We worked for two years. A great deal of this time was spent talking while seated face to face over a work table, standing at a nearby blackboard, or on the telephone. Many hours were spent in isolation, lost in thought over ideas and equations. We con-sulted with specialists in mathematics, cognitive science, neuro-biology, psychology, genetics, and the multifarious disciplines and subdisciplines of the social sciences, and we digested a li-brary of several thousand books and articles in these fields. We wrote lengthy memoranda to one another that served as the basis of further conversations and, eventually, the rough drafts of chapters of a book published in 1981, *Genes, Mind, and Culture*. Much of the work consisted of tedious probing, testing, and backtracking. Even so, less than a month after our first meeting we felt we had part of the answer and could see the way toward constructing an initial working theory.

The eidylons above possess a genetically programed culture, in contrast to the xenidrins below, who possess a genetically neutral culture.

# The Rules of
# Mental Development

IT WAS OUR BELIEF that the way into the problem was to try to envisage all the ways that a mind can evolve through millions of years and all the forms the mind might take in the end, at least within the limits of a broad classification. The human mind is only one of a vast number of possibilities that might have materialized here on Earth or elsewhere—on some other planet. By imagining these many possibilities, we can gain a better perspective on the one species available for study, making it easier not only to examine ourselves more closely but also, as Konrad Lorenz once put it, to look behind the mirror.

The Milky Way galaxy, of which humanity is a part, is composed of a hundred billion suns, roughly one for every nerve cell in the human brain. Farther out lie billions of other galaxies of almost unimaginable scope and variety. The laws of chance alone dictate that there is life around some of the stars, and because evolution is such an enormously creative force there are probably also advanced civilizations. In short, the universe might be as conceived in the poetic imagination of Jorge Luis Borges, an unending labyrinth of permutations where all possible arrangements eventually come to pass.

Imagine one such world. Sparkling cities of perfect geometric design, an athletic and alert citizenry, advanced forms of art and science, starships that travel to neighboring planets and suns, and a long history never marred by conflict: Utopia. But for the observer from Earth the great achievement stands forth as a deeply disturbing paradox. These aliens are complete genetic robots, even though they have created a civilization of high moral and artistic attainment. By considering the essence of such a genetic civilization we go beyond mere speculation to identify the deep relation between genes and culture. Let us call our imagined intelligent race the eidylons, from the Greek for "skilled ones." The full scientific name given them might be *Eidylus strictus*, or "rigid experts."

The eidylons are a brilliant and formidable species, but their entire thought and behavior are programed into their brains, right down to the exact words they use to piece together sentences. Language, art, and every other aspect of the culture are affected by the circumstances of the moment, but all are genetically predetermined in form. A ritual hymn is sung at a festival, stirring feelings of deep pleasure in the audience, but it is hereditarily fixed to the last note and inflection. In an atmosphere of intense excitement, a scientific team returns from a distant planet to report on its findings. The information is new, but the concepts and terminology used to describe it are genetically inherited and invariant. The eidylon civilization is transmitted across the generations according to the exact instructions of a particular set of genes. But, and here is the paradox that illumines the problem of mind and intelligence, the eidylons also teach their young all the details of eidylon culture. How can a civilization be simultaneously genetically fixed and transmitted by culture? The answer is that, although the eidylons teach and learn all that they know, they can transmit only one thing in each category—one language, one creation myth, one set of hymns, and so forth. Like the white-crowned sparrows of California (Earth), which must hear the song of their own species in order to learn it but are impervious to all else, they are unable to learn anything but the eidylonic repertory.

Mankind has traveled a radically different trajectory in its two-million-year ascent from *Homo habilis*. A defining trait of the human condition is the idiosyncrasy of the information that individual people learn, recall, and exploit. The sudden appearance of new opportunities, the identification of previously unsuspected rivals or allies, and unexpected disasters all intrude into personal lives to determine who will live and reproduce. The human environment contains much that changes quickly—a flux of history that would leave far behind individuals who were fully preprogramed by their genes. It was virtually predestined at the very beginning of evolution that people would not be automata. They would have to discriminate and select among the alternatives encountered repeatedly in the environment—that is, which hymn, which form of dress, which ethical decision. The decisions are individual, subject to peculiar differences in personal history, and prone to error. In this special sense human beings have free will. On Earth at least, genes and free will are partners of necessity and not partners of convenience.

The human trajectory was prescribed in part by the initial molecular properties of heredity in life on Earth. Because the genes of all organisms are made of nucleic acid (principally DNA, or deoxyribonucleic acid), of which only a certain amount can be stuffed into each cell, there is a strict limit on the number of genes that a human being or any other Earth-dwelling organism can possess. And this number greatly limits the complexity and flexibility of the mental processes that might otherwise be preprogramed in thought and communication. Perhaps a civilization could have been achieved with a much smaller, hard-wired mind. But would such a knowledge structure function effectively in the adaptive niche actually chosen by human beings? Mathematical estimates make it very doubtful.

So there is a kind of genetic destiny, one that nevertheless steered evolving mankind away from a frozen fate and toward the creation of free will. Augustine was wrong. We are not permanently flawed in the very way we were created, according to God's will. His great theological rival Pelagius was, on the other hand, more or less right. We are perfectible according to our own will. But the important question remains: did the genes create the mind only to free it totally from all antecedent biology? Were there "Promethean genes," which in effect liberated mankind from the remainder of human DNA?

In order to see what such a step must entail, let us make a second imaginary journey among the galaxies in order to search for beings who are the exact opposite of the eidylons. They are the xenidrins (*Xenidris anceps*), a species with true blank-slate minds. All cultural possibilities are equally open to the xenidrins. They can be taught any language, any song, any code of ethics, with equal ease. Their genes direct the construction of their body and brain but not their behavior. The xenidrin mind is entirely a product of the accidents of history, the place they live in, the foods they encounter, and the stray inventions of words and gestures. If we watch eidylons and xenidrins for short periods of time, their outward behavior might appear similar or even identical. But a close inspection of the growth of their children would reveal radical differences in the way their minds work: exquisite automata versus brilliant driftwood, iron will versus protean fidelity.

The burning question of human nature can now be put in simple form: are we closer to the eidylons or to the xenidrins? No one believes that we are even remotely like the genetically fixed eidylons, or could be like them, and scientific investigation has supported that conclusion. But a great many philosophers and social scientists do accept the opposite extreme view, that the mind is a blank state and man exclusively a product of his history and economic arrangements. In this xenidroid world there would be no human nature, only culture molding mankind. The anthropologist Leslie White said, "Culture exerts a powerful and overriding influence upon the biological organisms of *Homo sapiens*, submerging the neurological, anatomical, sensory, glandular, muscular, etc., differences among them to the point of insignificance." And so it follows that culture "changes and develops in accordance with laws of its own, not in obedience to man's desire or will. A science of culture would disclose the nature and direction of the culture process, but would not put into man's hands the power to control or direct its course." If this assertion is correct, the social sciences and humanities are totally independent of biology and the natural sciences, and must chart their own course. In fact, quite a few social theorists do think that this concept is about right and reject the idea that biology is relevant to the study of social behavior.

In our early theoretical calculations, however, we projected that it is no more likely for the human mind to be a slave of culture than it is to be a slave of the genes. There is a powerful tendency for the brain to evolve into a perpetually growing system that combines cultural innovation with genetic influence. In the end, when any intelligent DNA-based species such as *Homo sapiens* emerges, the individual mind must be able to reflect upon problems and make choices, but its growth and development are biologically programed to take certain directions in preference to others. We called this intermediate form of learning *gene-culture transmission*. To summarize the picture of all possible worlds briefly, there are three conceivable ways of transmitting culture from one member of the society to another.

*Pure genetic transmission.* The eidylon way. Although various choices exist, and individuals may be aware of them, only one can ever be preferred. Learning takes place but is rigidly channeled.

*Pure cultural transmission.* The xenidrin way. Multiple choices exist, and all are equally attractive and easily transmitted. The choices made by individuals depend entirely on culture and not at all on biological predispositions, which do not exist.

*Gene-culture transmission.* The human way. Although an immense array of possibilities can be learned, biological properties in the sense organs and brain make it more likely that certain choices will be preferred over others.

Even if a species could be created to resemble the xenidrins, with pure cultural transmission, evolution will eventually carry it from the blank slate and into a culture based on gene-culture transmission. We went on to calculate the average number of

## PURE GENETIC TRANSMISSION

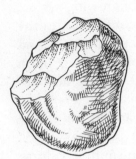

## PURE CULTURAL TRANSMISSION

## GENE/CULTURE TRANSMISSION

The three conceivable types of cultural transmission, illustrated by choice among hand axes and geometric patterns. The greater the preference for a particular alternative, the darker the image shown in this representation. The three types of ax are real (they are, left to right, Oldowan, Abbevillean, and Acheulean), but the preference depicted—for the middle, Abbevillean form—is speculative, as it might have existed in early *Homo erectus*. The reaction to the geometric patterns is real, based on Gerda Smets's experiments with adult subjects (see page 196 for details).

PURE GENETIC TRANSMISSION

PURE CULTURAL TRANSMISSION

GENE/CULTURE TRANSMISSION

generations needed to take a xenidrin-like species away from this extreme condition. We found that the time is reduced to just a few generations if there are many choices available.

The conception that began to emerge is that genes and culture are held together by an elastic but unbreakable leash. As culture surges forward by means of innovation and the introduction of new ideas and artifacts from the outside, it is constrained and directed to some extent by the genes. At the same time, the pressure of cultural innovation affects the survival of the genes and ultimately alters the strength and torque of the genetic leash.

To see how the linkage between genes and culture must arise, consider a blank-slate species, programed to make choices strictly according to cultural tradition, never influenced by inner biological urges or automatic procedures of thought. It is inevitable that some of the choices available to the society (say in diet, sexual behavior, or ways of counting) will confer greater survival and reproductive ability on the members who make one choice and spurn another. It is also inevitable that over a period of generations new genetic mutations and recombinations will arise that predispose individuals to make the adaptively superior choices. The new genetic types will spread through the population at the expense of the old. The species will evolve in such a way as to depart from the blank slate, and relatively soon compared to many other forms of genetic change.

Each member of the human species has to begin as a small child lost in a vast and complicated maze. From his starting point he must reach a place that represents an understanding of his culture and physical environment. At each turn of the maze, at each decision to learn and choose, he has only his wits and a list of clues to guide him. A premium is placed on speed and efficiency, because the child is racing against other children in similar mazes, and the greatest rewards of the contest will go to those who reach the exact goal with the least hesitation. In this contest, of which the child and the adults around him may be only dimly aware, there exists a tradeoff between wits and clues that can yield equal performance from the racers. Sharp wits, especially those designed to reason expertly about mazes, can do well with a short list of clues or a list that is distorted and confused. Dull wits, capable of only the most general kinds of reasoning, require a long, detailed list of clues in order to finish the race in equal time.

The maze is a metaphor for the problems facing a young child as he matures. The race is the image of evolution by means of natural selection. The growing mind must pick its way through a chaos of sensations and perceptions, quickly assembling them in a form that imparts a substantial degree of command over the environment into which it has been born. The entire course of the individual's life, including his physical survival and reproductive success, will be determined by the shape of the world that is recreated in the mind. In this contest, the blank-slate mind depends wholly upon the close guidance of a benevolent society.

In developing the models of gene-culture coevolution, we thought a great deal about the cultural environments in which children are actually raised, especially the living hunter-gatherer societies whose arrangements are most likely to resemble those of early man. We noticed a discrepancy between the data and a key assumption underlying much of the work on child learning, which holds that the social world of children is relatively well ordered and filled with instructions given by teaching adults. If this assumption were true, a child could afford to be dull-witted, in the sense of possessing only the most general and unconstrained learning abilities. But the actual facts are otherwise. In the Ituri Pygmies and !Kung Bushman of the Kalahari Desert, for example, children are left to their devices even more than in economically more advanced societies. They pick up most of their language and skills by voluntary imitation and apprenticeship. Remarkably enough, their minds flourish in these casual circumstances. They become sophisticated in speech, quotidian skills, and tribal lore. They also assume the facial expressions, taste preferences, and other general patterns of behavior that together form the diagnostic traits of human nature. These many qualities in the aggregate distinguish all of us from chimpanzees, eidylons, and xenidrins.

Behavioral scientists refer to the riddle of the self-taught child as the "principle of the poverty of the stimulus." In the imagery of the maze, the child's mind navigates correctly with a very short list of clues. How is this possible? Evolutionary reasoning suggests a straightforward answer: the genes supply much more than simply the general ability to solve problems. They equip the mind with specific rules and principles needed to learn the world quickly in an advantageous form.

Suppose that for thousands of generations young children varied in the way their minds grow. Some inherit only the most general of problem-solving mechanisms, while others are equipped with inborn clues and biases that accelerate mental development in certain directions. Most of the generalists end up in imaginary worlds poorly connected to the world they inhabit. In contrast, those provided with innate clues master the real social world into which they are born. They leave more and more of their genes in succeeding generations. The species evolves toward their type. In a literal sense the brain specialists, the swift and directed learners, will inherit the earth. Gene-culture transmission becomes the rule.

This inference about evolution raises the enormously important question of how much constraint actually exists in mental development. In approaching the problem we were careful to avoid thinking of any particular magnitude or to generalize across the whole range of human behavior. The time was overdue for a new and more interesting approach: to examine the development of human cognition and behavior category by category in order to discover the patterns of choice that actually exist. We asked: out of all the vast number of ways an intelligent species can evolve, which one has the human species actually followed during its evolutionary history? This is an empirical question, and fortunately it can be answered by observing human development directly. We began a tour of the technical literature on mental development, paying particular attention to data on choices made by infants and young children in circumstances that are relatively free of cultural influence. We also sought the advice of anthropologists and psychologists familiar with the subject.

We expected a windfall of information, but far less existed than we had hoped. The reason is that the emphasis in studies of child development has been on central tendencies, the "typical" pathways of development. Choices between the usual and the less usual are seldom recorded, even though most psychologists are aware of their existence. In the most celebrated of all studies of intellectual development, Jean Piaget charted the stages through which children gain increasing competence in abstract thinking and problem solving. But he and others in the Piagetian school seldom tried to examine the natural preferences of children for particular kinds of thinking and manual activity out of the large variety within their competence. Looked at another way, developmental psychologists working in this mode have made splendid progress in describing how typical human beings progress from infancy to the early stages of intellectual maturity, but they have not mapped the choices made at each twist and turn of individual development. Without such a comparative approach, Piagetian theory, as it is often called, is little more than a description of certain broad aspects of development. Although Piaget himself spoke of a genetic epistemology—the pathway in origin of knowledge—the true relation between genes and culture cannot be examined until optional trajectories of development are defined and some means devised to measure the biological bias influencing their selection.

The scarcity of useful information appears to stem in part from a peculiar and fundamental relation that has always existed between experiment and theory in science: the importance of experimental data is judged by the theory to which it is applied. As the physicist Arthur Eddington said half seriously, no fact should be accepted as true until it has been confirmed by theory. Unless an attractive theory exists that decrees certain kinds of information to be important, few scientists will set out to acquire the information. Also, the theory must be respectable and—even better—fashionable. No one had created a solid theory of human evolution that gives importance to the degree of biological influence on mental development. As a consequence, relatively few data were available to guide the construction of the theory.

Still we were able to locate published studies of twelve categories of behavior that contain sufficiently precise measurements of the mode of transmission. From this sample a remarkable result emerged: in every case the behavior is learned through gene-culture transmission; mental development appears to be genetically constrained. This result could not have been the result of observational bias. The psychologists who conducted the experiments were generally unaware of most of the other work being conducted of similar nature. They had no visible preconceptions about the mode of transmission; if anything, the Zeitgeist of contemporary psychology for the most part favors a belief in blank-slate minds. Yet the data from all the research programs revealed gene-culture transmission, a partial automatic preference on the part of the developing human mind for certain cultural choices over others. Some of the more striking examples produced by these pioneering studies entail the following familiar forms of thought and behavior.

• Only a very small percentage of individuals prefer to have sexual relations with brothers or sisters. They may harbor moments of inward desire toward siblings. But the vast majority choose to mate with persons raised outside their immediate family circle. Studies of the origin of sexual preference in Israeli kibbutzim and Taiwanese villages indicate that, even if other members of the society could somehow be neutral or favorable toward sibling incest, young people would still automatically avoid it in an overwhelming majority. The aversion is based on an unconscious process in mental development. Children raised closely together during the first six years of life feel little or no sexual attraction toward each other when they reach maturity, whether they are close relatives or not. As one anthropologist put it, people who use the same potty when very young do not marry when they grow up. The feeling has little to do with culture or the classification of kin. Even if a society could somehow begin anew with brother-sister incest as the norm, it would probably develop a cultural antagonism toward the practice in a generation or two. Eventually, the society would incorporate taboos in the form of rituals and mythic stories to justify and reinforce the aversion. In a phrase, the genetic leash pulls culture back into line.

• The learning of color vocabularies is also strongly biased and hence falls in the category of gene-culture transmission. From infancy onward, normally sighted individuals see variation in wavelength not as a continuously varying property of light (which it is) but as the four basic colors of blue, green, yellow, and red, along with various blends in the intermediate zones. This beautiful illusion is genetically programed into the visual apparatus and brain. Marc Bornstein at Princeton University used special techniques that measure attention span to show that four-month-old infants respond to variation in wavelength as if they were discriminating the four adult categories.

The same pattern occurs worldwide. At the University of California, Berkeley, Brent Berlin and Paul Kay worked with the native speakers of twenty languages, including Arabic, Bulgarian, Cantonese, Catalan, Hebrew, Ibibio, Thai, Tzeltal, and Urdu. The volunteers were asked to describe their color vocabulary in an unusually precise way: they were shown a large array of chips varying in color and brightness, and directed to place each of the principal color terms of their language on the chips that came closest to their conception of what the words mean. Even though the words differed strikingly from one language to the next in origin and sound, they fell into clusters on the array that correspond, at least approximately, to the principal colors distinguished by Bornstein's infants.

The physiological basis of the partitioning in vision is partially known. The color cones of the retina, which are the cells that distinguish wavelength, are differentiated into three types that approach but do not correspond exactly to the basic colors. These cells are maximally sensitive to blue (440 nanometers), green (535 nanometers), and yellow-green (565

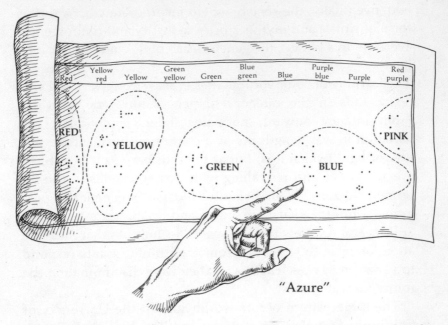

The role of biology in the formation of color vocabularies. Each point on the color chart represents the position of a color term in one or the other of twenty languages. The points tend to cluster in positions corresponding to the four basic colors perceived by the human brain.

nanometers) respectively. In the lateral geniculate body of the thalamus, one of the key relay stations between the eye and the visual cortex of the brain, the visually active nerve cells are divided into four types that appear to encode the principal hues. The deeper mechanisms that translate these diverse sensitivities into the conscious perception of color are under active investigation. Few brain scientists doubt that a full explanation of color vision at the levels of the cell and molecule will eventually become possible. Furthermore, simple genetic changes in color vision, creating the various forms of color blindness, occur widely through human populations. They have been associated tentatively with the malfunction of particular genes located on the X-chromosome.

The intensity of the learning bias was strikingly revealed by an experiment conducted on color perception during the late 1960s by Eleanor Rosch of the University of California at Berkeley. In looking for "natural categories" of cognition, Rosch exploited the fact that the Dani people of New Guinea have no words to denote color; they speak only of "mili" (roughly, dark) and "mola" (light). Rosch considered the following question: if Dani adults set out to learn a color vocabulary, would they do so more readily if the color terms correspond to the principal innate hues? In other words, would cultural innovation be channeled to some extent by the innate genetic constraints? Rosch divided 68 volunteer Dani men into two groups. She taught one a series of newly invented color terms placed on the principal hue categories of the array (blue, green, yellow, red), where most of the natural vocabularies of other cultures are located. She taught a second group of Dani men a series of new terms placed off center, away from the main clusters formed by other languages. The first group of volunteers, following the "natural" propensities of color perception, learned about twice as quickly as those given the competing, less natural color terms. They also selected these terms more readily when allowed a choice.

• Infants prefer to look at objects that have particular shapes and arrangements, and as time passes their choices change in a predictable manner. From birth they gaze longest at pictures that are large, contain numerous elements, and consist of curved lines. Most of all they favor figures whose outlines contain approximately ten independent turns. By the age of eight weeks they also prefer bull's-eye designs over parallel stripes, touching elements over those that are separated, and irregular arrays of elements over those that are perfectly aligned. These apparently innate biases parallel an early preference for the abstract design of a normally composed human face over various humanlike but scrambled designs. By twenty weeks the infant shifts its attention increasingly to

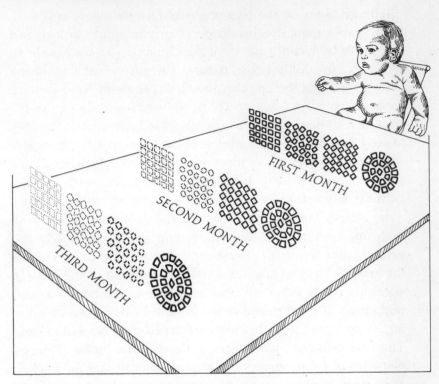

Infants prefer certain geometric designs over others, as indicated by the duration of their gaze. The degree of the preference changes with age.

new designs and faces in preference to those it has already learned, and as a result its visual experience expands rapidly.

• Although facial expressions vary from one culture to the next, strong tendencies exist that must be classified as gene-culture transmission as opposed to a purely cultural form of transmission. People around the world use a common set of expressions to register fear, loathing, anger, surprise, and happiness. Paul Ekman of the University of California at San Francisco tested the strength of this predisposition in an elegant manner. He photographed Americans acting out these emotions and New Guinea highland tribesmen as they told stories in which similar feelings were emphasized. When individuals from each culture (New Guinea or American) were then shown portraits from the other culture, they interpreted the meanings of the facial expressions with more than 80 percent accuracy. This was the case even though the New Guinea tribesmen had been previously exposed very little to the outside world, while the Americans who looked at the pictures knew nothing of the Papuan culture.

The distinctive nature of the brain's program in facial recognition is further illustrated by the rare medical condition called prosopagnosia. When lesions occur on particular regions of the undersurface of the temporal and occipital lobes of the brain, the patient cannot identify other persons by their faces. In extreme cases he is unable to recognize the features of even his closest relatives. The disability is not due to a general loss of visual memory; the patient can still identify objects other than faces by sight alone. Nor is it due to an inability to remember different people; the patient can distinguish them by their voices. The bizarre properties of prosopagnosia demonstrate how the brain can be biologically programed to follow specific sensory cues, especally when the category of learning is concerned with the most pressing needs of social life.

• Newborn infants choose most kinds of sugars over plain water and in this descending order of preference: sucrose, fructose, lactose, and glucose. They also discriminate among substances that are acid, salty, and bitter, reacting by twisting their faces into the characteristic adult expressions of distaste for each substance. This selectivity continues into childhood and has important effects in the evolution of adult cuisines.

• Anxiety in the presence of strangers occurs in very young children in all the many cultures around the world studied by the German ethologist Irenäus Eibl-Eibesfeldt. The baby turns away, buries its face in its mother's shoulder, and often begins to cry. This relatively complicated response first appears at six to eight months of age and peaks sometime during the subsequent year. It does not depend on previous unpleasant experience with strangers; nor does it appear to be linked to crying and other signs of discomfort caused by separation from the mother. The latter development is distinct in appearance and first emerges when the infant is about fifteen weeks old. Anxiety in the presence of strangers continues at a lower, controlled level into childhood and even maturity. It slides easily into fear and hostility, contributing to the tendency of people to live in small groups of intimates. These responses are intensified when strangers stare. Eyes and eye-like patterns have been found to have a generally higher arousal effect on people of all ages than do other facial features. They are also key elements in the attraction of newborn infants to the face as opposed to other parts of the body, and they play a central role in communication afterwards.

• The innate tendency for human beings to learn one thing as opposed to another, in other words gene-culture transmission, is perhaps most dramatically illustrated by the phobias. These are the extreme fears into which people are plunged—stricken by nausea, cold sweat, and other reactions of the autonomic nervous system. Phobias typically emerge full-blown after only a single unpleasant experience, and they are exceptionally difficult to eradicate, even when the victim is carefully reassured and coached by a psychiatrist. It is remarkable that the phobias are most easily evoked by many of the greatest dangers of mankind's ancient environment, including closed spaces, heights, thunderstorms, running water, snakes, and spiders. Of equal significance, phobias are rarely evoked by the greatest dangers of modern technological society, including guns, knives, automobiles, explosives, and electric sockets. Nothing could better illustrate the peculiar and occasionally obsolete rules by which the human mind is assembled, or the slowness of man to adapt to the dangers created by his own technological triumphs.

For convenience we decided to label the various regularities of development as *epigenetic rules*. Epigenesis is a biological term that means the sum of all the interactions between the genes and the environment that create the distinctive traits of an organism. Thus the color vocabulary used by a person is based on the interaction of genes prescribing color perception in his eyes and brain with the environment in which he developed. This environment ranges from the fetal conditions that produced his eyes and brain to his subsequent enculturation. The epigenetic rules of color vision and classification are stringent enough to direct cultures around the world toward the central clusters of color classification as revealed by the Berlin-Kay experiments. But they are not strong enough to impose completely identical classifications on every culture and every person.

The epigenetic rules of mental development are a menagerie of diverse but still largely unstudied phenomena. During the past twenty years, psychologists and brain scientists have uncovered evidence of developmental regularities in even the most subtle and complex forms of mental activity. People follow unexpected and sometimes remarkably inefficient procedures in the way they recall information, judge the merits of other people, estimate risk, and plan strategy. Among the peculiarities of decision making is the excessive use of stereotypes. When observers are asked to guess the occupation of another person who is shy, helpful, and obsessed with detail, they are more likely to choose librarian over other occupations, even when their personal experience runs counter to this conclusion. Most people, including some trained statisticians, intuitively expect small random samples to reflect faithfully the large population from which they are drawn, although this is demonstrably untrue in a large percentage of the cases. Other studies have revealed that human beings are also poor intuitive statisticians when dealing with the major events of life and death. They tend to equate events that have a low probability and low consequence with events that have a low probability but important consequence. As a result they underestimate the effects of catastrophes. In particular, they consistently misjudge the future effects of warfare, as well as floods, windstorms, droughts, and volcanic eruptions, even when such events are repeatedly experienced and remembered over many generations. Other examples of developmental bias in language formation, logic, and basic arithmetic will be described in the next chapter, when the evolutionary origin of the modern mind is more fully examined.

Most of the mental idiosyncrasies of the human species have not been translated by psychologists into a precise calculus of cultural choice. But the evidence accumulated so far indicates clearly that the genes have not freed the mind in the special sense of conferring pure cultural transmission. They hold thought and culture in an intermediate degree of dependency, and that, we felt, might contain the secret of the accelerated evolution of the human brain.

As we pieced together the scattered data, we became convinced that this way of viewing cultural transmission, if put in systematic form and appropriately applied to psychological data, could provide more complete models of mental evolution than had been conceived previously. Not only did the formulation seem to solve the nature/nurture dilemma in a form more satisfactory than any other we had encountered, but it also offered the means of devising an exact analysis of the interaction between genes and culture.

Between genes and culture! When one tries to envision the bridge across that enormous gulf all at once, the imagination fails. The full chain of causation runs circuitously up from DNA through the formation of the nerve tissues and the glands that secrete hormones. The exact configurations of the brain cells and the manner in which the hormones affect them are the chief determinants of the epigenetic rules. The rules then shape the outcome of culture.

By means of painstaking research, reported in a hundred journals, scientists have begun to trace this complicated sequence. They have provided convincing examples of the way that the brain is assembled under the direction of the genes. To take a graphic example, locomotion is modulated through relay centers in the cerebellum, the three-lobe knot of nerve tissues located at the rear of the brain. In mice a series of rare recessive genes have been discovered that alter the fine structure of the cerebellum. When a mouse possesses a double dose of one of these mutant genes rather than a single dose or none at all, the arrangement of the principal layers of nerve cells in the cerebellum is changed during embryonic development in ways that deviate profoundly from the normal condition. Other mutant genes alter the stucture of portions of the inner ear and other regions of the brain that contribute to coordination. Depending on which of the genes is present, the mice suffer various bizarre alterations in their movement. The changes have been officially christened by geneticists as waltzing, twirling, reeling, trembling, staggering, agitans, zig-zagging, quaking, pirouetting, waddling, and fidgeting.

Comparable mutations, in different categories of cognition and in greatly varying degrees of sublety, have been discovered in human beings. Sex is determined by a single pair of chromosomes out of the twenty-three pairs comprising the full complement of human hereditary material. If the fetus receives two X-chromosomes it becomes a female, while the possession of an X- and a Y-chromosome makes it a male. On rare occasions an embryo ends up with a single X-chromosome, and the result is a female who will always be short in stature and sexually immature. Women with the deficiency also typically possess high verbal ability but very low capacity for spatial reasoning, such as following directions on maps.

The pinpointing of such effects provides the means for an ultimate "genetic dissection" of the brain and behavior, in which the many processes of development are distinguished according to the differing sets of genes that control them. For example, a gene has recently been discovered that alters performance on three standard psychological tests of spatial ability but not on three other tests of spatial ability and seven tests that measure verbal facility, perceptual speed, and memory. More such surgically precise mutations are likely to be found as the genetic search widens. By 1980 no fewer than 3100 human genes had been identified out of a total of perhaps 50,000 belonging to the primary, "structural" category. Of these, 340 had been placed on one or the other pair of the 23 pairs of chromosomes. Hundreds are known to affect behavior at least indirectly.

It may be disconcerting at first to think of a single gene that alters behavior in some narrow, specific manner—at least, for such a gene to influence *human* behavior. Many thoughtful people object to the entire conception of a linkage between genes and culture at just this point: surely *I*—the inner self, with my powerful volition and free flights of creative thought—*I* am not in any real sense the product of a bunch of genes. This legitimate concern can be dealt with by reentering the subject through anatomy and the rudiments of behavior. Whether you can roll up the sides of your tongue to form a tube, or whether the tongue remains flat and quivering with such an effort, is determined by a single gene. Now extend your thumb back as far as possible; whether the upper digit bends back 45 degrees from the base digit ("hitchhiker's thumb"), or whether it turns outward gradually and at a lesser angle, is determined by a single gene. Lay your hand palm down on the table and spread your fingers out; whether the little finger is perfectly straight or curves gently in toward the other fingers is determined by a single gene.

From such elementary cues we can proceed to others that are more "psychological." Single genes have recently been discovered that turn on or off the ability to smell particular musks and possibly also floral scents. Even volition can be affected. The Lesch-Nyhan syndrome, a rare condition controlled by a single gene, causes violently aggressive behavior: the afflicted people curse uncontrollably, strike out at others around them with no provocation, and even chew and tear at their own lips and fingers.

Actually, hundreds or thousands of genes are involved in the manufacture and maintenance of every organ or pattern of behavior. But chemical variation in any one of them can cause a variant such as hitchhiker's thumb or the Lesch-Nyhan syndrome. By detecting and analyzing large numbers of such single-gene differences, great and small, a picture of the full genetic blueprint can gradually be assembled. This is the way genetics has proceeded from the garden plots of Mendel to the mighty enterprise it is today.

We are at the dawn of human behavioral genetics, when the first efforts are being made to discover and characterize the genes that guide the development of more complex forms of behavior and thought. Consider the facial expressions that denote the basic emotions of fear, loathing, anger, surprise, and happiness. They appear at an early age with little variation across all cultures. There is an excellent possibility that mutants of facial expression exist but have gone undetected because of lack of attention from scientists. It would not be surprising to find genes that predispose individuals toward poker faces, excessive smiling, or other, more disconcerting facial tics.

To go further and grasp the full relation between genes and thought, it is necessary first to establish the philosophically important point that mind has a material basis. Throughout the history of ideas two fundamentally opposed theories have competed for dominance: dualism, which holds that mind is a nonphysical substance created by the brain but existing apart from it; and materialism, which considers the mind to be an exclusively physical activity of the brain. Dualism, like the similar conception of the *élan vital* (nonphysical essence of life), has steadily given ground to the contrary results of experimental studies and is now almost extinct. It has failed in good part because no one has been able to conceive of a way by which a noncorporeal force can set muscles into action and guide behavior— at least not without repealing the laws of physics. Even so, one celebrated neurophysiologist, John Eccles, has run against the tide by postulating not two but three levels of reality: the world of physical objects, the world of mental activity, and the world of man's culture and knowledge. But Eccles' ontological scheme is not very convincing. It is based less on evidence than on his personal hopes for the separate existence of a pure ego and a spiritual nature by which the individual "transcends the evolutionary origin of his body and brain."

The materialist philosophy of mind has found its most satisfying modern form in what has been called central-state identity theory. The basic proposition is that mental events are identical with physiological events in the brain, and most probably the coded pattern by which particular sets of nerve cells are electrically discharged. The greatest advantage of this quite common-sensical idea is that a physical basis is allowed for pure mental activity. Thus the binding restrictions of behaviorism are set aside, and for the first time since William James examined such old-fashioned notions at the turn of the century, research on the innermost processes of reflection and feeling have become fully respectable. Central-state identity theory has been further strengthened during the past fifteen years by "functionalism," a philosophy of mind that examines mental process not just as the activity of sets of nerve cells but as an embodiment of *any* information-processing device, whether nerve cells, silicon chips, or even some unimagined form of intelligence waiting to communicate with us from Arcturus. Functionalist theoreticians have drawn their best insights from the study of computer design and machine intelligence, and they stress software over hardware— the details of programing rather than the physical form of the circuits. They conceive of psychological intention not as a puzzle posed for the delectation of philosophers but as a real procedure that can be programed into the nerve cells of the brain, either by human hands or by organic evolution. Once psychological intention has been solved as an engineering problem, the scientist's reach should extend to explanations of predisposition, drive, and even mood. The central problem then becomes the set of computer programs that most closely simulate human mental activity. The closeness of the match between the behavior of artificial brains and the behavior of human brains will serve as an acid test of the materialistic view of the mind.

Many will consider this conception the ultimate conceit of science. Not so long ago skeptics easily dismissed the whole approach by arguing that, since the physical basis of mental activity cannot be seen directly, the mind-body problem can never be the

subject of more than inspired guesswork. Yet even this formidable technical barrier has started to break down. Using improved
techniques, researchers can now monitor electrical activity over
the entire surface of the brain, filtering out moment-to-moment
variations while tracing changes in the patterns of activity that
occur with age and mental disease. An alternative technique is to
measure the flow of blood near the surface of different parts of
the brain. Xenon 133, a mildly radioactive form of the inert gas
xenon, is injected into one of the main arteries in the course of
medical diagnosis. As the patient performs various physical and
mental operations, the increase of blood flow to the surface of
various sectors of the cortex is detected as a rise in gamma radiation emitted by the xenon 133. Still deeper probes into the living
brain have become possible with an arcane technique called PET
(Positron Emission Tomography). Chemicals normally used by
the brain in metabolism are first labeled with the radioactive isotope flourine-18, each atom of which replaces a hydrogen atom
in the metabolic chemicals. When flourine-18 decays it emits
positrons (positively charged electrons), and the positrons give
off gamma rays when they combine with electrons. The gamma
rays are then picked up by detectors outside the body and their
main points of origin are triangulated with the aid of computers
(this scanning procedure is called tomography). Although PET
still has low resolution, it has been used to locate activity within
the visual processing areas of the brain when subjects think
about pictures. It has also been possible to place the recollection
of music even in the absence of sound. People who tried to remember a particular melody activated portions of the right hemisphere, while those who visualized the notes on a musical staff
activated the left hemisphere.

The mind springs from a machinery of neurons created according to the genetic blueprint, but it grows in an environment
created by the preexisting culture, which is a particular history
embedded in the memories and archives of those who transmit
it. Consequently the way in which memory is implanted and organized is crucial to the creation of culture. Psychologists have

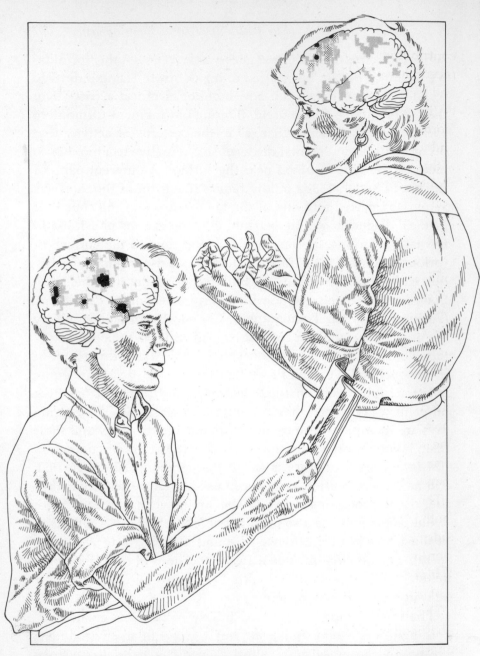

The physical location of mental events. Silent reading and counting are associated with the activation of particular regions of the brain, as depicted here by darkened patches on the cerebral cortex. These results were made possible with a technique that uses the radioactive isotope xenon 133; the degree of shading represents the amount of local physiological activity as indicated by the intensity of radiation.

contributed substantially to the understanding of this subject in recent years. They have made a basic distinction between short- and long-term memory. Short-term memory is the "ready state" of the conscious mind. The mind can handle only about seven words or other symbols simultaneously. It takes about one second to scan them fully and forgets most of the information within thirty seconds. Long-term memory takes much longer to acquire, but it has an almost unlimited capacity, and a large fraction of it is retained for life. The conscious mind summons information from the store of long-term memory and holds it for a brief interval in short-term memory. During this time it processes the information—at a rate of about one symbol per 25 milliseconds—in order to make decisions.

Psychologists have further identified two forms of long-term memory. Episodic memory recalls specific events by drawing particular persons, objects, and actions into the conscious mind through a time sequence. Thus it easily replays the image of a runner leaping a hurdle. In contrast, semantic memory creates *meaning* in the form of related concepts simultaneously experienced. Thus fire is likely to be connected to hot, red, dangerous, cooked, and so on; a sharp cry evokes apprehension or hostility. The concepts are the nodes or reference points in long-term memory. Many concepts are labeled by words in ordinary language, but others are not.

The two forms of long-term memory are closely related to one another. Semantic memory originates during real episodes, while nodes of semantic memory—such as the scent of a rose or the sound of a barking dog—often evoke the memories of past episodes that consist of these and other nodes arranged into a particular time sequence.

The nodes of long-term memory are almost always linked to other nodes, so that the recall of one node summons certain others into the conscious mind. A link may fit one or the other of several categories recognized by psychologists. Ascriptive links attach particular properties to an object or action (hence swift to runner, violent to boxer, elegant to pas de deux). Denotative

The formation of culture from node-link structures in long-term memory. The !Kung hunters associate different kinds of animals and events with traits and names, represented in the diagram as circles connected by straight lines. They also make links to emotional feelings, which are symbolized by wavy lines. After building such node-link structures through experience (*left*), they reassemble them as a time sequence to recall episodes and to tell stories (*right*).

links call up a word or some other symbol, while emotional links evoke associations to feelings that are typically "difficult to put into words."

Long-term memory is thus based upon what some psychologists have appropriately chosen to call node-link structures. These complexes comprise the ideas, criteria, and plans of action from which the mind is created. In an influential study published in 1967, M. Ross Quillian proposed a "spreading-activation" theory of long-term memory involving node-link structures. His initial purpose was to construct a superior computer-search technique based on a model of the brain, but later he and other investigators transformed the basic technique into a more explicitly psychological theory. In essence, the theory says that the brain learns by constructing a growing network of concepts. When a new entity is experienced, say a song, a fruit, or a mathematical technique, it is processed by a spreading search through the network in an attempt to find links with previously established nodes. A novel form of fruit is immediately classified by its shape, color, texture, palatability, and the circumstances in which it was encountered. It is linked not just with other kinds of fruits but also with a whole repertory of emotional feelings, memories of dietary custom, and recollections of previous but similar discoveries. In terms of the spreading-activation theory, the mind is visualized to a substantial extent as an enormously complicated and constantly changing network of nodes and links. Research has scarcely begun on the way long-term memory is stored, especially at the level of nerve cells and physiology. But in the meantime the node-link structure is a powerful metaphor for summarizing much of what is known.

We now come back to the epigenetic rules, which are the regularities in mental development that lead to a preference of one mental operation over another. Each of the rules can be expressed as a strong tendency to form particular node-link structures. Familiar examples include the remarkably swift bonding of mother and infant, the specificity of phobias, the preference of newborn infants for sugar, and the strong and predictable association of eye movements with particular kinds of facial communication. The brain finds it easier to store some nodes in long-term memory, and in addition it links them with certain preexisting nodes far more readily than with others. This is the manner in which the epigenetic rules guide the assembly of the mind. The regularities in development result in what is outwardly perceived as the most general traits of "human nature." The human genes prescribe the epigenetic rules, which channel behavior toward the characteristic human forms of thought.

We became convinced that the epigenetic rules as understood by developmental psychologists are consistent with the properties of node-link structures as understood by specialists on memory. When these two main components, the epigenetic rules and the node-link structures, are combined, they provide a picture of mental development sufficient for the immediate purposes of theory. Of course, the model falls far short of incorporating the full complexity of mental life. Of spirit, dreams, and epiphanies it says very little. But it does capture enough of memory and emotional reward to permit the analysis of culture as the collective knowledge and beliefs of a society.

Thinkers traditionally stress that the hallmark of human evolution is plasticity: man the creative animal, man freed from the savage bonds, the spirit struggling to rise above its baser nature. Like all capsular generalizations that are both reasonable and pleasing, this one is a half truth. Gene-culture coevolution expanded the human mind but did not, could not, and perhaps *should* not fully emancipate it. During its pell-mell race to the *Homo sapiens* level, the human species incorporated quirks and biases into most and possibly all categories of cognition and behavior. Mental capacities and emotional responses were shaped in ways that favored particular levels of inquisitiveness and optimism, as well as idiosyncratic modes of reasoning, a stubborn tendency to see the world as a peculiar medley of lights, sounds, and smells, and the deep, emotionally controlled preference for certain cultural choices over others. The result with which human beings live today is the general occurrence of an intermediate degree of bias in their mental development. People are neither genetically determined nor culturally determined. They are something in between, a much more interesting circumstance.

We have argued that a blank-slate mind, using pure cultural transmission, is a near impossiblity in the evolution of any intelligent species. The brain that evolved toward this extreme state would be a purposeless calculating machine with a limited capacity for survival. If the evolution of man had somehow reached such an end point, there would be no human nature, no wells of passion, nothing truly distinctive about the way people think and feel except the algorithms programed into them by outside and independently acting forces. The eighteenth-century philosopher David Hume said that reason is the slave of the passions, and then he added, with considerable prescience, that this is the way it should be. In other words, human nature and not pure reason is what lifts us above the inanimate world.

During our early conversations at Harvard in 1979, our reasoning consisted of the construction of "plausibility arguments"—ideas that describe nature and solve problems only in very general terms. We felt that the theory of gene-culture coevolution, if conceived correctly, might account for the principal features of human evolution. It predicted a rapid increase in the size of the brain, the general occurrence of the intermediate (gene-culture) form of cultural transmission, and the existence of cultural diversity that varies greatly from one behavioral category to the next.

But plausibility arguments are only the beginning of science. Now we faced a tougher question: what could such ideas *do*? Conventional social theory, consisting of such formidable enterprises as Marxism, structuralism, and psychoanalytic theory, is also based on plausibility arguments. Its often elaborate constructions are internally consistent. They convince and excite people, and they have even changed history. But this does not mean that they are true. The harder task for social theory is to confront human nature in the same way the natural sciences confront the physical world, and thereby to transform plausible scenarios into concrete phenomena and exact predictions. The more detailed the results of the predictive models, the more vulnerable they are to testing, and the more convincing will be their parent theory if they survive.

The theory of human nature that prevails in the end will be the one that aligns social behavior and history with all that is known about human biology. It will correctly and uniquely characterize the known operations of the human mind and the patterns of cultural diversity. That is the grail toward which many scholars toil, despite frustration and not infrequent humiliation. At issue are the very limits of the natural sciences. Is the quest a fool's errand? Are the social sciences destined to remain an independent domain of inquiry? Those trying to join biology to the social sciences can easily be defeated by the still unmeasured technical problems, while opening themselves to charges of trespassing, incompetence, and presumption. But the prospects are so attractive that the risk seems worth taking. The time is ripe, as Peter Reynolds has said, to stare biology square in the face and start all over again.

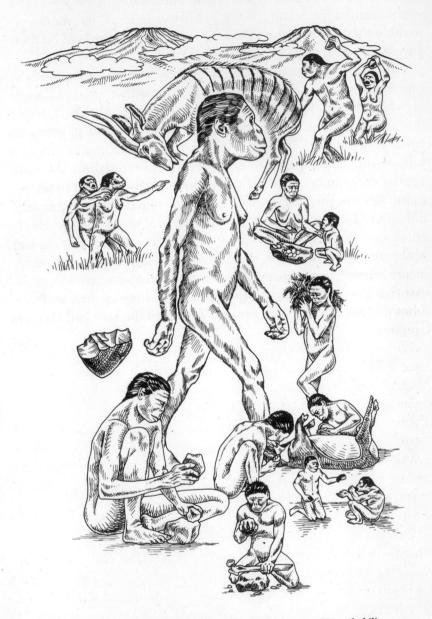

A speculative view of the world of the earliest "true" man, *Homo habilis*.

# The Social Worlds
# of *Homo*

THE UNIQUE QUALITIES of the human mind evolved in the two-million-year period during which *Homo habilis* metamorphosed into modern man. The essential product—the luxuriant epigenetic rules guiding mental development—is only now receiving the full attention of scientists. *Homo habilis,* the earliest true man, must have also possessed a distinctive set of rules, although simpler in design. If we could somehow recreate the sequence of events that led from the process of mental development in *Homo habilis* to that of modern *Homo sapiens,* we would penetrate closer to the origin of the human condition. To do so it would be necessary to proceed in steps through time from one set of epigenetic rules to another, just as paleontologists dig through rock strata to uncover fossils and deduce evolutionary trees.

Of course it is difficult enough to decipher the evolution of human anatomy from a few fragments of bone. It is a far more daunting and less certain task to reconstruct the history of mental events from the relatively tenuous evidence of developmental psychology. But the task is not hopeless, and the question is much too important to abandon without carefully weighing every datum and clue. If scientists have ventured onto shaky

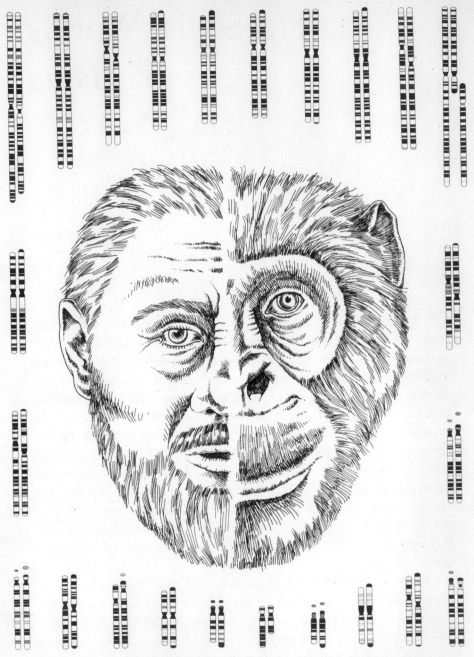

The recent common ancestry of human beings and chimpanzees is suggested by the close similarity of their chromosomes, the carriers of the genes. There are twenty-four kinds of chromosomes in each cell (22 plus the $x$ and $y$ chromosomes, which together determine sex). The human chromosome is the one on the left; the chimpanzee counterpart is immediately to its right. (Chromosome diagrams: copyright 1980 by the American Association for the Advancement of Science.)

ground in this domain, they nevertheless have very good reason to press forward—treading softly.

Specialists in evolutionary biology reconstruct a vanished species with the aid of a straightforward and proven technique. They *bracket* it between the species to which it gave rise and other, still more primitive living species. Several lines of biological evidence suggest that man descended from a primitive African apelike species that also gave rise to the chimpanzee. *Homo sapiens* (man) and *Pan troglodytes* (chimpanzee) are in fact so similar in details of anatomy, physiology, chromosome structure, and enzyme chemistry as to resemble many pairs of animal species in such groups as fruit flies and birds that are definitely known to have split from each other within the past million years. Researchers believe that the chimpanzees and earliest man-apes originated from the same ancestor five to twenty million years ago, so that a common origin of these two lines is entirely within reason. The bracketing technique is applied in this case in the following manner. Those traits found in both the chimpanzee and modern man are also considered likely to have existed in *Homo habilis*. The evolutionary tree is believed to form a rough Y, with the several key species occupying these positions:

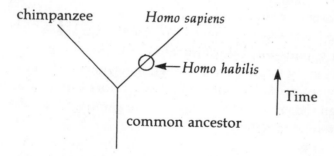

It follows that traits shared by the chimpanzee and modern man (*Homo sapiens*) were probably also shared by both the common ancestor and *Homo habilis*, the intermediate species between the common ancestor and modern man. The overlap in traits does not constitute proof (few forms of evidence in evolutionary biology ever do), but it can be taken to indicate a fairly high level of likelihood that the traits occurred over the entire length of the two arms of the Y. The behavior and mental activity of a vanished species can be recalled, at least in part.

What are these common traits? We have woven them together, along with the evidence about food habits and campsites deduced from fossil remains, to produce the following speculative portrait of the world of *Homo habilis*.

As DAWN SPREADS from behind the Ngorongoro peak in East Africa, a dozen habiline men, women, and children climb westward down a dry stream bed toward the lakeside camp of a larger band. The encounter between the two groups is tense. Some of the opposing adults recognize each other, but the last meeting was months before and memories are dim. The older males advance. Some stare tight-lipped, a universal primate sign of challenge; others purse their lips in an expression of growing excitement. A dominant male of the larger band turns openly hostile. He motions the opposing band away by swinging his right arm forward as if throwing an object underhand. But the gesture is ignored. First the adults and then the young mingle. They extend their hands and touch each other's bodies lightly, talking back and forth with conciliatory murmuring. The individual sounds are full of meaning but are probably not true words—not arbitrary symbols strung together to form sentences. As anxiety subsides, smiling and short bursts of laughter spread through the group. The young start to gambol, chase each other, and pretend to fight.

The larger band has surplus meat. On the previous day the foraging males discovered the fresh corpse of a young hippopotamus sprawled in the mud of a stream-bed hollow. Several ran off to gather handfuls of large basaltic pebbles for conversion into tools. One of the habilines selected a stone six centimeters in diameter, around which the hand could be curled easily, for use as a hammerstone. He struck it three or four times against other

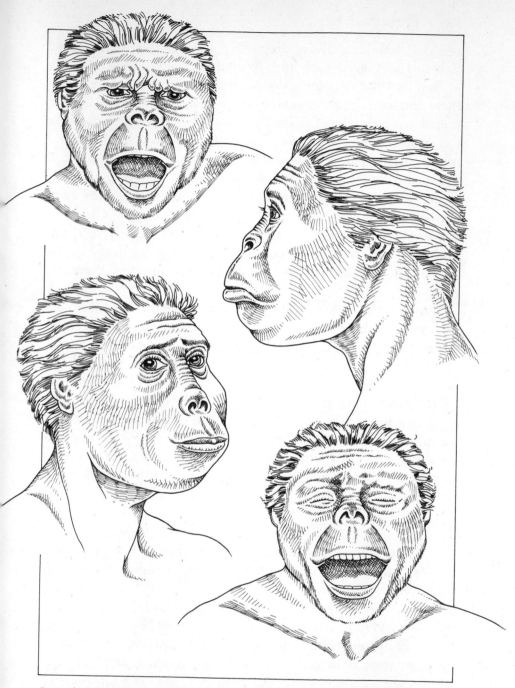

Some facial expressions of *Homo habilis,* inferred from those that are similar in chimpanzees and modern human beings. From the top: anger, pouting, repose, laughter.

pebbles, breaking away irregularly shaped but sharp-edged flakes. The pieces were then pulled back and forth as little knives, and the remaining cores became crude choppers. With these implements the hunters cut into the hippopotamus and pulled away chunks of flesh. They chopped open the brain case to reach the brain, cracked the rib cage to uncover the heart and lungs, smashed the long bones to expose the marrow, and sliced the thick-skinned anterior abdominal wall to lay bare the kidneys, liver, and spleen. They ate the choicest parts on the spot, then sawed the tough, fibrous joint ligaments to free a limb for transport back to the home base. The stone tools were abandoned with the remains of the carcass, to await discovery by the archaeologist descendants of the hunters one hundred thousand generations later.

Now, on the following day, the larger band shares its meat with the members of the smaller one. The excitement the food engenders is intense, and the aroused habilines communicate eloquently with rapid gestures and vocal signals. A few of the newcomers are able to eat by unobtrusively pulling fragments off the pieces held in the hands and mouths of others. But most must beg for shares, by holding out their upturned palms and whimpering. That afternoon, stirred by the restless pacing and hand pointing of its largest male, the smaller band departs for a distant forest enclave. There are no farewells; none yet exists in the human vocabulary.


FOOD SHARING, conducted in rudimentary form by chimpanzees and carried to extreme ritualistic heights by modern man, was probably crucial to the social life of the habilines and as much as any single trait of anatomy marked the beginning of their long evolutionary ascent. Anthropologists believe that *Homo habilis*, unlike any other primate with the possible exception of the man-apes, both gathered food *and* carried it for long distances. The habilines were also unique in the large amounts of animal food they collected through hunting and scavenging.

They enjoyed an exceptionally broad diet. From time to time, perhaps daily, groups of foragers left the campsites in search of fruits, berries, nuts, tubers, and meat. Some of this food must have been eaten on the spot, but a portion was brought back to the camp to be shared with those band members who remained behind. The beneficiaries almost certainly included crippled adults and mothers with infants. From the pattern of wear on the stone choppers, it is known that the habilines used their tools to cut both large tubers and animal flesh. It seems likely that adults of both sexes participated in foraging, but that the males traveled farther and concentrated more on hunting. Such is the division of labor employed by virtually all living hunter-gatherer societies. And in this instance the habilines are nicely bracketed on the other side of the Y-shaped evolutionary tree: when chimpanzee groups hunt young baboons and monkeys, the lead is usually taken by the males.

Although the habilines bequeathed us only their bones, meager traces of food, and the circles of rocks marking their home bases, they must have also employed a more complicated "soft" industry. Chimpanzees are remarkably ingenious at inventing and using perishable tools. They crumple leaves and sponge up water from tree holes, and strip leaves from twigs and fish for termites by poking the twigs into the nests. They also uproot saplings to lash enemies and employ twigs as toothpicks. The habilines, with larger brains and the undisputed capacity to shape stones, almost certainly possessed at least that varied a repertory. Glynn Isaac, a leading authority on the reconstruction of Paleolithic environments, has pointed out that the habits of hunting and carrying food were powerful stimuli for the invention of other simple tools that might easily have been contrived by intelligent apes. He believes that the most primitive humans used sticks to spear animals and dig soil, and that they transported food in turtle shells, bark trays, and stomach bladders.

Chimpanzees have other, sometimes surprising talents. Under laboratory conditions they can weave sticks and vines into simple patterns (but cannot untie knots). They can classify and group objects into abstract classes according to size and color, distinguish photographs of human beings from those of all kinds of animals, and draw rough circles and other elementary figures just short of representational images. When a chimpanzee looks into a mirror he recognizes himself as something distinct from other members of his own species. In the original test of that capacity, the psychologist Gordon G. Gallup put spots of red dye on the heads of chimpanzees under anesthesia and then allowed them to see their reflections after awakening. The apes immediately responded by touching their hand to the red spot. We may conclude that if some habiline Narcissus ever looked into a pool of still water, he understood that the face staring back was his own image and not that of a second, ghostly primitive. Perhaps he also thought in some wordless fashion: *this is I*, who exists apart from the clamorous band and will someday die. Scientists, given enough time, might deduce whether this is true and thereby have something to say about the evolutionary history of the self and of the soul.

Biologists and psychologists alike speak of flexibility as an advanced trait and, sure enough, chimpanzees and great apes have more varied behavior than monkeys. When given a toy or some other novel object to examine, they touch it with more of their body parts, hold and manipulate it in a greater variety of ways, and are generally less predictable in moment-to-moment responses. As a corollary, young chimpanzees play and explore more than other animals, yet much less than modern human children and adults. We can again assume that the problematic habilines lay somewhere in between. Play extends the variability of behavior mightily and opens numerous possibilities for cultural innovation in both animals and man. John and Janice Baldwin described a remarkable example involving a two-year-old squirrel monkey named Corwin. Occasionally Corwin dropped food pellets, which bounced off his cage floor. He turned the accident into a game in which he deliberately dropped pellets and chased them as they bounced around. One day as he was leaping upward a pellet flew out of his hand and ricocheted through the upper part of the cage before settling to the floor. Corwin then started to release pellets deliberately as he jumped, making the game more complicated. Finally, he learned to toss the pellets up into the air and catch them in his mouth.

Such antics can sometimes be turned to advantage. One of the subordinate male chimpanzees studied by Jane Goodall at the Gombe Stream National Park in Tanzania learned to bang two empty kerosene cans together. He then used the extraordinary movement and noise to augment his threat displays and, as a result, rose to dominance in just a few days over larger males in the troop. Another, partially crippled chimpanzee observed by Geza Teleki compensated for his lack of mobility during hunting by dashing the head of a prey repeatedly against tree trunks. How easy it would be to evolve to a more humanlike behavior, to change from hitting a stick with a head to hitting a head with a stick. The habilines or their immediate ancestors almost certainly took this step. They inaugurated the long and malevolent lineage of weaponry, which in its final nuclear form could annihilate *Homo* and demonstrate—in a conclusive and unexpected manner—that culture is indeed superior to heredity.

Of the chimpanzee epigenetic rules, those processes by which the ape's mind is assembled step by step, we know almost nothing. The course of chimpanzee intellectual development has been charted to some extent by psychologists who have extended conventional work on humans to include these apes. As a result we have an increasingly clear picture of how well chimp infants can manipulate objects. We also know the age at which growing youngsters can solve elementary puzzles, memorize the meaning of symbols, and practice primitive forms of art. But tests on choice, by which apes are allowed to pick among flavors in drinks, geometric designs, ways of holding objects, facial expressions, and so forth, have not been undertaken. And little wonder: the importance of such analyses for basic theory are still largely unappreciated even in the case of human beings. It is clear that epigenetic rules of mental development do exist in these animals and that they are subject to bias just as in human beings. Chimpanzees avoid incest in a humanlike manner, consistently rejecting as sexual partners those band members with whom they were most closely associated as juveniles. They also have at least one kind of response that outwardly resembles a human phobia. When chimpanzees were shown a stuffed leopard in the African wilds, their reaction was explosive. They dashed about barking and shouting, hugged and kissed each other, and voided their bowels. Some broke off saplings to lash the "monster," and finally, when the harmlessness of the stuffed animal became apparent, the troop closed in to inspect it with what can only be categorized as awe.

These accounts of course rank as no more than suggestive anecdotes. But they also point the way to definitive experiments. It should be relatively easy to compare the responses of the apes to similar, competing stimuli—for example to leopards as opposed to lions and hyenas—through successive ages and degrees of previous experience. A pattern of cognitive development will emerge, and when the information is matched point by point with the rapidly accumulating data on human mental development, the bracketing technique can be used to sketch a tentative picture of the mind of *Homo habilis*. Then it will be possible to assess with greater confidence the events that occurred during the ascent from the habilines to modern man.

What we have in the interim is a substantial mass of information on the way of life of *Homo erectus*, the transitional human species that arose from *Homo habilis* about 1.5 million years ago and gave rise to *Homo sapiens* a  million years later.

The fragments of bone and stone tools that *Homo erectus* left behind have led archeologists to form a portrait of a human being more intelligent than the apes and *Homo habilis*. The *erectus* bands followed complicated patterns of migration through base camps and temporary resting sites, while pursuing different game according to season. To exploit the environment through such a long-term rotation must have required improvements in memory, foresight, and leadership. What is more, for such skinny, fangless bipeds to hunt animals as large as elephants, in

fact just to seize the fallen bodies from other scavengers, required impressive skills. The elephants might have simply blun-
dered into swamps and bogged down in the syrupy mud. But it is
just as likely that they were driven into such natural traps. How
could a small group of human beings herd panicked and very
dangerous animals ten times their weight? A good answer is fire.
Bits of charcoal are distributed around the ancient hunting sites
in a pattern suggesting that *Homo erectus* set fires that swept over
large areas of grass and brush. "My guess," the anthropologist
F. Clark Howell has written, "is that the purpose was to drive
elephants along the valley into the swamps." We are evidently in
the presence of a creature far above the level of the most intelligent animals, one that can be called human in a fuller sense.
Much of the available information can be summarized in another
fictional scenario.


A *Homo erectus* toolmaker squats on a stony ridge and
searches the terrain below him. Nearby he sees the rising smoke
of the campfire and the hurrying figures of the newly arrived
band members as they drag in tree branches for firewood and
shelter. He listens to a familiar cacophony: the cracking of wood,
shouts, laughter, a steady murmur of primitive speech—vocal
signals delivered emotionally, perhaps a scattering of true words.
His eyes quarter along the more distant terrain, south across an
expanse of wind-stunted pines and copses to the deep-blue arc
of the Mediterranean Sea. He thinks briefly about what may lie
beyond, on the other side. No hope of ever knowing. It is mid-
morning, one million years ago.
  The band has been on the move for days. How many the toolmaker does not know and cannot conceive. His days are hour-
less, he has no concept of years, and higher numbers are forever
beyond him. But like all of his kind, he senses the change of time
in the daily passage of the sun and stars, in the seasonal cycles of
the grass and wildflowers and movement of game. These matters
he knows very well, and in them he is wise even by modern

The world of *Homo erectus*, the species of man intermediate between habiline and modern man: a speculative reconstruction.

human standards. They fill his thoughts now, after a sleepless night of dread.

Early in the evening before, as the light failed, the band was visited by a pride of hunting lions. The lionesses circled the human bivouac briefly. They crouched facing the huddled band, their tails switching lightly from side to side, the telltale intention movements of the hungry predator. Then inexplicably they rose in unison and left. But they did not travel far. Through the night the band heard an occasional deep cough, a rustle of disturbed shrubs and fallen branches here and there. The lions were still hunting, now deterred by the smoldering campfire. The human watchers very likely wondered what else might be out there moving in the night. Perhaps the strange stumptailed cats with saberlike teeth, a vicious pack of hyenas, and other, formless horrors belonging more to the imagination than to reality, the forerunners of monsters and bogeymen. Better to have a generalized fear of the dark and to shrink thrilled and apprehensive from the unknown than to take time to learn and deal with each menace in turn. On this night, deep in *Homo erectus* time, men occupy a paradoxical position within the ecosystem. Their tools and organized movements are turning them into the greatest predators of all time; yet their thin, slow bodies render them prey for the most powerful of their carnivore rivals.

The toolmaker picks up a hammerstone in his right hand. It is a quartzite pebble slightly larger than a clenched fist, tapered to a blunt edge. Extending his left hand he selects a round unworked stone of equal size. As he makes these first movements a group of children halt a game of king-of-the-mountain and climb the slope to watch. The toolmaker hefts and turns the two stones in his hands. He is judging, choosing, thinking of the finished product, why it is needed, to whom it will be given, and how it will be used. His mind at this instant is a flurry of competing possibilities. He settles on a sequence of steps. Concepts of vision and sound crowd through his consciousness in the form of a time series, like labeled beads sliding along a string. Perhaps he links silent words with the concepts that pass in review, so as to say roughly: "Strike . . . turn . . . edge . . . ax . . . give . . . brother . . . horse." He will dress an ax to give to his brother for the butchering of a horse. This Ur-language, if summoned, is a poor accompaniment to the rich and fluid imagery of the concepts upon which, in the course of human evolution, they are being grafted. A stunning linguistic efflorescence will ensue sometime during the next million years.

The toolmaker holds the rough stone in his upturned palm and grips it tightly with his fingers. He pulls his arm close to the chest, tenses, and strikes down hard with the hammerstone. A chip flies to the side, leaving a concave depression and a sharp ridge along the thin edge of the stone. In the next hour the toolmaker repeats the process fifty times: turn and examine, grip, and strike. With shorter, more precise blows he trims the opposite sides into a double edge. In the end an almond-shaped hand ax emerges. It is a fine example of the Acheulean industry, the mark of *Homo erectus* culture, elegant in comparison with the crude choppers of *Homo habilis* yet still far inferior to the splendid stone instruments later made by *Homo sapiens.* For some reason the toolmaker chooses to stop at around fifty percussions instead of proceeding to the hundred or more required for a significantly more refined tool. Perhaps he cannot conceive of anything better. His most ambitious fantasies cannot reach much further than the tool he now cups in his hands.

The toolmaker descends to the camp, trailing his entourage of children. The simple bowl-shaped shelters have been completed. A group of foragers is leaving at this moment to search for tubers, berries, and small animals, and with any luck to sight big game. There will be a sharing of food and exchange of tools.

In several days the band will move on toward a winter rendezvous point. There they will join a friendly group, composed of familiar faces. Some of the adults will be recognized generically as kin. In the ancient hominoid manner there will be an exchange of young females. The communication will be intense, an emotional mixture of sounds and quick gestures. There may also be some true words—short, distinctive sounds used as symbols and conveying arbitrarily chosen meaning.

Some of the men will organize long-distance marches to hunt big game. Their parties will move with growing circumspection as they approach a distant stream that runs to the sea. There lies the territorial boundaries of strangers. During vicious raids those aliens have killed members of the band, and a few have fallen in turn. Although they look the same (and are in fact *Homo erectus*), these creatures seem wicked and not truly human. They are almost as little known as the unseen forces patrolling the night. If they could be destroyed or driven away, the entire band would experience indescribable relief and joy. There would be an urge to stripe the body with ochre and to dance.

SO THE HUMAN MIND was growing genetically more complex during the million-year transitional period of *Homo erectus*. But although *erectus* evolution was quick relative to that of most other organisms, it was glacially slow in comparison with the acceleration that later carried *Homo sapiens* from the Paleolithic era to the beginnings of civilization. In some localities there was even a deterioration rather than an improvement in toolmaking skills over periods of a few hundred thousand years. These anomalies can be explained by recognizing that the whole *erectus* population, which extended all the way from Africa to eastern Asia, was composed of thousands of groups isolated from one another by rivers, mountain ranges, and sheer distance. The bands probably contained no more than thirty or forty members, and of those only several were likely to have been skilled toolmakers. If the best craftsman died in a hunting accident, the abilities of the group as a whole might easily have been set back for years.

The size of the *Homo erectus* brain nevertheless grew centimeter by centimeter—it eased its way upward for a million years. As the anthropologist Ralph L. Holloway has pointed out, it is a mistake to attribute this evolution solely to the use of tools. The sophistication of the tool industry of Paleolithic man was probably no more than a crude index of the full power of his mental development. Far more important were the intricate social rela-

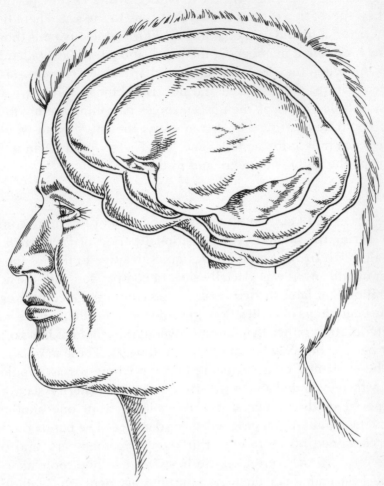

The growth of the brain through the three successive species of human beings: *Homo habilis* to *Homo erectus* to *Homo sapiens*. The forms of the two smaller brains (belonging to *habilis* and *erectus*) are based on casts made of the interior of fossil skulls.

tionships being woven during the *Homo erectus* period. Today even the simplest of hunter-gatherer and agricultural societies, employing relatively crude lithic, wood, and bone implements, match or exceed groupings of stockbrokers and college professors in the complexity of their social arrangements. Their daily existence is a choreography of talk and posturing governed by the rules of kinship, rank, taboo, and obligation. Punctilio may well have been the dancing master that directed the main part of the evolution of mental development.

Half a million years ago the earliest definable *Homo sapiens* came upon the scene as *Homo erectus* writ large, with a substantially more voluminous brain and a growing capacity for language. About 40,000 years ago the evolution of the material culture began to accelerate in a spectacular manner. Representational art appeared in the form of clay and stone sculpture along with simple but often strikingly beautiful paintings on cave walls. In some parts of Europe an elementary form of literacy had become well established by 32,000 years ago. It consisted of scratches on ornaments, pieces of bone and clay, and stones. The scratches were arranged through repeated motifs into descriptive classes such as meanders, fishlike images, and parallel lines. All these inventions appeared near the end of the anatomical evolution of *Homo sapiens.* Some anthropologists, including Clive Gamble of the University of Southampton and Alexander Marshack of Harvard, believe that man possessed a high level of intelligence during most of the half-million-year period but that some special environmental requirement triggered the use of the latent abilities only after a long delay. The fossil record indicates that the start of the artistic phase coincided with an unusual rapid growth and expansion of the human populations in the northern hemisphere. The symbols and representational art may have been called into play to register membership, status, and position. But in this tangle of events, which is cause and which is effect? The circumstances of the late Paleolithic cultural explosion remain one of the key mysteries of human evolution.

Three capacities separate modern man from animals: extreme

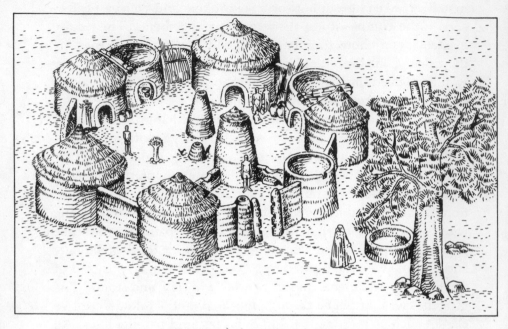

The social world of contemporary man is very complicated even in relatively "primitive" societies. A typical example is the patrilineal household of the Tallensi tribe of Ghana. Outside the compound depicted in this generalized drawing, the sacred tree and conical ancestor shrine mark the area reserved for male social gatherings. Near the gate the patriarch stands in the cattle yard, which is reserved as his special domain. His ancestral spirits dwell in the nearby flat-topped cattle shed. The granary directly behind him is also under his exclusive control. Proceeding clockwise around the homestead, to the left of the gate, are found a room for adolescent boys; the bedroom, pantry, and roofless kitchen of the patriarch's mother; and the living quarters of the patriarch, his wife, and his children.

powers of long-term memory, symbolic thinking, and language. The first of these characteristics must be immediately qualified. Human memory is in reality highly specialized. People can learn symbols and sentences immeasurably better than the most intelligent animals, including chimpanzees. But they are far from being geniuses in kinesthetic memory, the learning of pure sound and movement. Humpback whales, which may be the animal champions in this category, can perform an individualized "song" lasting over thirty minutes, and repeat it note for note many times, a feat that would strain the talents of an opera star. Human beings are also relatively mediocre at remembering space and time. Guided by only a few training runs, honeybee workers can recall the exact location of as many as five beds of flowers as well as the time of day each bed yields nectar.

Some behavioral scientists believe that the rudiments of symbolic thinking preceded true language. Chimpanzees and orangutans can recognize the concept of a piece of fruit from its odor or touch and transfer that bit of thought to a visual image of the fruit. We employ "thought" loosely in this case. The apes appear to be doing just what human beings do when they smell an orange or apple and point to an image of the fruit, but there is no way of picturing the concept in their minds. Chimpanzees can also learn a word such as "orange" and associate it with the identifying smell, touch, or image of the fruit. They can be taught short sentences, and once in a while they invent a new word or phrase. But even with the most conscientious human guidance they cannot create significant new concepts or grammatically consistent sentences. At best, language training appears to give chimpanzees a new means of expressing what they already know.

There are other major differences. Young chimpanzees have a drive to communicate, but it is of a special, limited kind. They engage in what psychologists call gestural babbling, a free wheeling invention and testing of body postures, hand signals, and facial expressions, along with some use of sound. Over a period of months this chaotic mixture grows into the characteristic

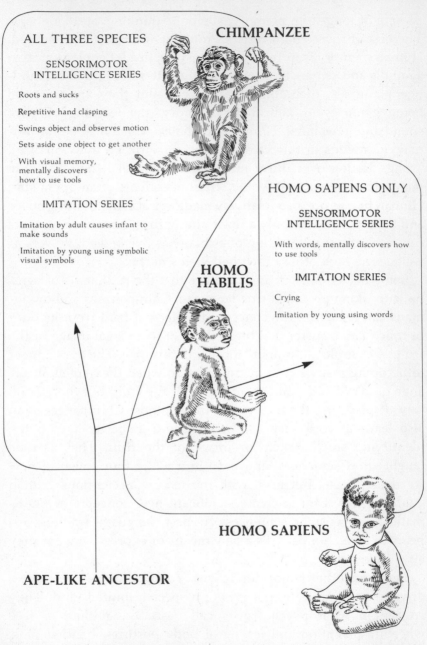

**ALL THREE SPECIES**

SENSORIMOTOR
INTELLIGENCE SERIES

Roots and sucks

Repetitive hand clasping

Swings object and observes motion

Sets aside one object to get another

With visual memory,
mentally discovers
how to use tools

IMITATION SERIES

Imitation by adult causes infant to
make sounds

Imitation by young using symbolic
visual symbols

**CHIMPANZEE**

**HOMO SAPIENS ONLY**

SENSORIMOTOR
INTELLIGENCE SERIES

With words, mentally discovers how
to use tools

IMITATION SERIES

Crying

Imitation by young using words

**HOMO
HABILIS**

**HOMO SAPIENS**

**APE-LIKE ANCESTOR**

The stages of mental development in chimpanzees and modern human beings, with those held in common and hence inferred to have occurred also in *Homo habilis,* the ancestral human species.

adult repertory, plus a few personal idiosyncrasies. Human children, in contrast, engage in *verbal* babbling. From an early age they seem driven from within to create and test new words, meanings, and sentences. So characteristic is this inventive stage that its product is recognized as a special child's language, with peculiar qualities of rhythm and syntax. Even the mere beginnings of the child's language pushes far beyond the mature ape's ultimate capacity. And where the ape must be urged and guided into the use of language, the human child often must be forced to shut up. Furthermore, the child's mind is equipped with epigenetic rules that lead it quickly to the choice of certain grammatic rules over others. They guide the mind to the more complex operations of sentence formation and thought. Thus the child's language matures easily into the adult tongue.

During the million-year transition from *Homo habilis* (wordless? apish in thought?) to *Homo sapiens*, the neocortex of the brain, containing the principal centers of association and thought, more than doubled in size. Although tracings from the inside of fossil skulls are inconclusive, profound architectural changes must also have occurred. One neural domain that did expand visibly was Broca's area of the frontal lobe, a major organizing center of langauge. When Broca's area is severely damaged through a head injury, patients retain their ability to understand individual words and phrases but experience great difficulty in constructing sentences. Their speech is telegraphic and ungrammatical. When Wernicke's area, located to the rear of Broca's area on the parietal lobe, is damaged, the opposite effect follows: patients can still speak grammatically correct sentences but their words are mostly empty of meaning. They make frequent mistakes such as substituting "chair" for "table" without understanding what's wrong. These clues from medical research suggest that the unique human capacity for language did not evolve through the random piling up of more neurons in the cortex of the forebrain. Language is not just the inevitable spinoff of a generalized intelligence. It is the peculiar product of a recently created division of labor among specialized portions of the brain and novel epigenetic rules.

The world of the modern human species, *Homo sapiens,* with its vastly increased culture and array of artifacts.

*Sapientization* is a word used by students of human evolution to label the essential changes that led to *Homo sapiens* from *Homo erectus*. The exact timetable for this final step is still unknown. The process began sometime during the long tenure of *erectus* and might even have roots in the evolution of the ancestral *Homo habilis*. But on the final products of sapientization there can be little disagreement: language, symbolic thinking, and the deepening of long-term memory to store the vast requisite information. Experts on classification have linked the three species of *Homo* together on the basis of anatomical distinctions, a sharing of certain properties of dentition and jaw structure, and the enlarged brain with its expanded association areas. They have clustered the man-apes together as a second genus, *Australopithecus*, and linked them with *Homo* to form the family Hominidae. This arrangement is a useful aid for remembering anatomical similarities. It is also an absurd distortion. If the epigenetic rules and outward behavioral traits were used instead of gross anatomy, and then if the probable microscopic distinctions in brain structure were added, *Homo sapiens* would almost certainly be ranked as a taxonomic family by itself. *Homo habilis* and *Homo erectus* might also be justifiably separated from the man-apes as a second family.

The succession of three whole mammalian families within three or four million years, as suggested in this arrangement, represents an extremely swift passage of evolution. One of the pivotal events was the invention of words, sounds that summon concepts of long-term memory. The concepts are in turn the knowledge structures of the mind by which images are formed and shuffled to expand the stream of consciousness. The traditional wisdom of linguistics is that language originated in the form of pure symbolism and was added to the bestial sounds and gestures of the ancestral prehumans. Animal communication is fundamentally iconic; it imitates the object or action that is intended or desired. The nestling songbird begs food by thrusting its head up and opening its beak wide, while the rhesus monkey threatens by staring and slapping its hands on the ground. More sophisticated signals evolve in animal communication by the process of ritualization, during which the original movement lose their original function and become increasingly conspicuous and stereotyped. The male of the European cormorant, a crow-sized fishing bird, courts the females by flashing his wings and raising his head in a garish imitation of the takeoff leap. African chameleons defend their territories with a striking display in which the sides of the body are pumped in and out in exaggerated respiratory movements. At the same time, the head is wagged and jerked in ritualized thrusts. Often the ritualization process has been carried so far that the original function is no longer clear. In such cases the communication superficially resembles the pure symbolism that supposedly characterizes all human speech.

Mary LeCron Foster, a comparative linguist, has proposed that human language originated by a ritualization process of the movements of the mouth and tongue and of primitive sounds. Like many contemporary students of human evolution, she believes that true human language is only about 50,000 years old and arose in conjunction with art and the rapid evolution of the materials-based culture. Although modern languages change over a period of only a few thousand years in the construction of their words and phrasing, the basic one-syllable sounds composing the words change much less rapidly. By examining the branches of the great Indo-European family of languages (Germanic, Italic, Hellenic, Slavic, Indo-Iranian), linguists have deduced the ancestral tongue, proto-Indo-European, that was spoken in the third millennium B.C. A family tree of the words exists, and the most persistent sounds can be identified. These primitive elements do appear to be at least partially iconic. In the minds of the Paleolithic inventors of language, the way the mouth is moved and the parts of the air tract utilized might have been directly linked to the meaning of the sound. For example, the *m* sound is articulated by pressing the lips tightly together while air is forced through the nasal passage and the vocal cords are vibrated. This sound is used in words across many languages to denote surfaces that touch, press, and hold together; words

A Cro-Magnon woman forms the sound "em" as a direct representation of the act of clasping. This example is based on the speculative reconstruction of the early evolution of language.

that mean crushing or resting against; and words that refer to fasting, chewing, and swallowing. Familiar examples include *mouth* (English), *mano* (hand in Spanish), and *main* (hand in French). Sounds created well back of the lips and nose, in other words with the tongue, teeth, or alveolar ridge, typically have more internal meanings, such as the experiencing of emotion.

As languages evolved, most words lost their direct representational content and became progressively ritualized until their iconic beginnings were entirely erased, just as in the evolution of the most advanced forms of animal communication. At that point the resemblance stops. Where animals are locked into the particular repertory of their species, note by note and gesture by gesture, human beings invent signals and freight them with arbitrary meaning. The primitive languages of the Paleolithic hunter-gatherers might have been conjured from iconic visions of the mouth and vocal apparatus. But an act of will, a mere whimsy, can produce a neologism that severs the ties with the archaic sounds.

Even so, the considerable maneuvering room of the human mind does not leave it entirely free to create any language with equal ease. The mind is guided by epigenetic rules, powerful enough so that the combined languages of the world, and probably all the languages that *could* evolve with any ease in the future, form only a microscopic subset of the verbal languages that can be imagined. The rules forbid entry into most of this theoretical domain. The child-in-the-maze, to return to our original metaphor, is guided far more quickly to the adult tongue than could be the case if every sound and grammatical rule were equally easy to learn.

Since Noam Chomsky argued the evidence for the existence of innate grammatical rules in the late 1950s, researchers have made substantial progress in identifying such constraints and charting their effects on language. Much of the information accumulated by the linguists is technical in nature, but it boils down to the special ways in which children acquire knowledge about their own language and transfer it into the coherent sentences of speech ("transformational grammar"). These procedures are the equivalent of what we have been calling the epigenetic rules of mental development, with special reference to the ways words are strung together to create meaning.

The constraints on language are linked to even deeper rules that affect our conception of reality. Philosophers use the word *ontology* to refer to beliefs concerning what can and cannot exist. Recently this most abstract of all inquiries has been turned into a scientific pursuit. Psychologists such as Frank C. Keil have begun to investigate the precise mental steps by which the human perception of reality is created. They have focused their attention on the predicates of language—those propositions that appear sensible, are automatically acceptable, and hence are used to build up long-term memory. For example, the predicate "is honest" may be applied to Napoleon. Whether or not the emperor was in fact honest, and whether all the time or just once in a while, the proposition at least makes sense. Beneath that predicate can be placed subordinate predicates that seem to follow as consequences or special conditions, such as "is reliable," "is sincere," and so on. In contrast, the predicate "is geometric" is not permissible. It cannot be linked at any level with the "is honest" hierarchy without violating the sense of reality. This leads to the M-constraint of human thought, first proposed by the philosopher Fred Sommers. Stated concisely, the rule is that two predicates cannot be linked to three terms if each predicate has a term that makes sense for it alone. Thus the following combination is forbidden:

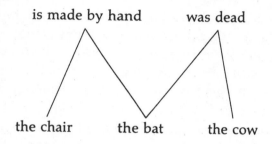

Faced with such intolerable ambiguity, the mind rushes to sever the M-structure. It breaks up the predicates and terms to form alternative branching trees that do not take the M-shape. When Keil tested 300 English- and Spanish-speaking children, he found that almost all had the M-constraint. As people grow older, they build longer and more complex trees of predicates that serve as the scaffolding of their thought. Without understanding the process explicitly, they follow the M-constraint.

In some way that neither philosophers nor scientists have fathomed, the adoption of discrete symbols has also led to deep constraints in the way quantity is envisioned; and from this symbolization of quantity, a human mathematics has been formulated. From the earliest age at which such matters can be examined, about two and a half years, children acquire four regularities in counting from which they never deviate the rest of their lives:

- *The isomorphism rule:* when counting a group of objects, one and only one number (or, more precisely, enumeration tag) may be assigned to each object.
- *The stable-ordering rule:* numbers must be applied in an order that does not change from one count to another.
- *The cardinality rule:* the last enumeration tag in a counting series signifies a cardinal number representing the quantity of the objects or process measured.
- *The order-irrelevance rule:* the assignment of objects to numbers, although fixed for the direction of a count, is arbitrary; in other words, the order of enumeration is irrelevant.

Other intelligent species may not necessarily think of quantity in just the same way as human beings. They are especially likely to differ during the earlier stages of their mental evolution. To take one of a vast series of imaginary cases, consider the eidylons during the hunter-gatherer period of their evolution. Counting is performed only under strict conditions and as part of a ritual. The rules are genetically fixed (recall that the eidylons can transmit only one culture). As hunters bring in game, the prey are counted by the shamans according to the following procedure: the dead animals are enumerated in strict order according to size, with the number 1 tagging the largest, the number 2 denoting the next largest, and so on. Thus the order-irrelevance rule of human beings is violated. If two animals appear to be the same size, they receive the same number, violating the isomorphism rule. In order to serve the deep religious belief of the unity of nature, a group of the smallest prey are set aside and counted in reverse order, violating the cardinality rule. The arithmetic of the primitive eidylons is built solidly into their epigenetic rules of mental development; they would have great difficulty coping with the human way. Nevertheless, their procedures are highly advantageous for them, just as ours are for us.

This fantasy shows that it is not at all difficult to imagine species of mind elsewhere in the universe that work very differently from the human mind and with greater or lesser efficiency. The human brain is constructed to handle language, reality, and quantity in a special human way and not in some other. The fact that much of this particularity appears self-evident to the very minds operating within its constraints does not diminish the sense of wonder we should feel by distancing ourselves from our idiosyncrasy long enough to see it as a product of organic evolution.

The origin of the epigenetic rules of higher reasoning was sapientization itself: it constituted the final stage of human evolution, which may have occurred as recently as the past fifty to one hundred thousand years. The purely *Homo sapiens* epigenetic rules were added to the surviving processes and regularities of mental development that had evolved in *Homo habilis*, *Homo erectus*, and still earlier ancestral species. This fundamental base includes incest avoidance, color perception, the patterns of mother-infant bonding, facial expressions, and the other forms of cognition and social behavior described earlier.

The modern mind is a biological palimpsest. Over thousands of generations the ancient script of epigenetic rules was partially erased to make room for the new. By careful examination of the sometimes jumbled pieces, using techniques from several scientific disciplines, it should be possible both to read the contemporary message and to reconstruct the history that produced it.

The coevolutionary circuit of genes and culture.

# Promethean Fire

So WE RETURN to the major question, which directs us to the Promethean mystery. Given that epigenetic rules occur at all levels of mental development, and given that culture forms the most important part of the human environment, what is the exact manner in which heredity and culture are connected?

The genes and culture are inseverably linked. Changes in one inevitably force changes in the other, resulting in what we have termed gene-culture coevolution. This uniquely human process is an enchanted circle into which the species has been drawn and will travel for the rest of its existence. The generations repeat the following sequence endlessly:

- The genes prescribe the rules of development (the epigenetic rules) by which the individual mind is assembled.
- The mind grows by absorbing parts of the culture already in existence.
- The culture is created anew in each generation by the summed decisions and innovations of all the members of the society.
- Some individuals possess epigenetic rules enabling them to survive and reproduce better in the contemporary culture than other individuals.

• The more successful epigenetic rules spread through the population, along with the genes that encode them; in other words, the population evolves genetically.

In sum, culture is created and shaped by biological processes while the biological processes are simultaneously altered in response to cultural change.

GENE-CULTURE COEVOLUTION might have begun this way in *Homo habilis.* Our distant ancestor had a brain somewhat larger than that of the modern chimpanzee. The limited archaeological evidence suggests that it occupied campsites for lengthy periods and constructed the most elementary forms of stone implements. It had reached a sufficient level of intelligence to remember a few more things, a bit longer, than chimpanzees. Perhaps it had acquired the earliest version of true language by investing arbitrary meaning in some of its primitive sounds and gestures. Thus *Homo habilis,* while still partially simian by contemporary human standards, had already inched past the chimpanzee level of culture. If this interpretation is correct, habiline man had crossed a threshold of enormous importance. For the first time in the history of life, the coevolutionary mechanism was activated.

The primitive *Homo* now experimented with language and social behavior. At first the genes held such probing in close rein, permitting the assembly of only a limited memory and a minuscule vocabulary. Music, art, and myth lay far in the future; they were beyond the capacity of the habiline mind or even any predisposition to invent them. But at an early stage the cultural reconnaissance must have yielded the ability to communicate basic information on subjects that were most important to survival and reproduction: the location of feeding places and water holes; the key traits of neighboring bands, whether friendly or hostile, strong or weak; and the degrees of kinship and behavior of fellow band members.

As this capacity expanded over thousands of generations,

*Homo habilis* evolved into the wide-ranging *Homo erectus.* The primitive cultures, facilitated by the growing capacity to remember, spread slowly into new realms of language and social behavior. They pulled on the genetic leash and began to change it. With a greater genetic capacity for intelligence came a wider array of choices in perception, reflection, and behavior. But, as we have seen, this intelligence was not machinelike; it assumed a specialized, distinctively hominid character. When individual band members made certain choices in preference to others, say the use of stone implements, they enjoyed a higher rate of survival and reproduction. So those who possessed epigenetic rules directing them to the right choices were better represented in later generations. Genes prescribing the most efficient epigenetic rules spread through the population over many generations. Human populations evolved toward those forms of learned behavior that conferred the greatest survival and reproductive power.

Gene-culture coevolution as we conceive it is nicely illustrated in elementary form by the case of incest, and especially brother-sister incest, which has been closely studied in several aspects. During mental development the individual makes a choice between two kinds of sexual partners: siblings versus nonsiblings. Those who choose siblings have more defective children because of the effects of inbreeding. The epigenetic rules that turn the mind away from incest lead also to cultural patterns that reinforce the rules—taboos and frightening mythological stories. The coevolutionary process is set in motion: persons who conform to the aversion and taboos leave more healthy offspring; the genes underwriting the avoidance of incest remain at a high level in the population; and the predisposition is sustained as one of the epigenetic rules. The two inheritance systems have evolved in concert—the genes that create the epigenetic rules and the culture that offers the choices on which the rules act. All of the events are linked into a circuit of causation, running from the genes to the rules of mental development to culture and back to the genes again.

The circuit can be more fully visualized as a spiral stretching out across time. Its exact form is the trajectory of human evolution. However, as important as the details of the trajectory are for their own sake, their examination can be postponed for the moment. It is the underlying process of gene-culture coevolution, rather than isolated events in the history of the species, that created the human mind. To trace the full circle of gene-culture coevolution is to understand the unique qualities of human evolution.

We believe that the origin of mind will be fully disclosed only when three challenges are met. The first is to explore the coevolutionary circuit through its most accessible part, from the workings of the individual mind up to culture. The second is to proceed back along the circuit, from culture down to the genes and the individual mind. The final challenge is to explain why man alone entered the circuit.

## The First Challenge: Mind to Culture

The crucial first step is to examine the process that leads from the making of individual decisions to the formation of culture. At this level men manufacture their own history. Free will appears to be in charge. As Engels observed, "Each person follows his own consciously desired end, and it is precisely the resultant of these many wills operating in different directions and of their manifold effects upon the outer world that constitutes history." We approached the mind-to-culture translation with the aid of a proven rule of scientific research: theories are most easily constructed and understood if they manipulate discrete elements. Atoms, molecules, and genes are examples of such units that appear to be real and discrete. Populations, species, and ecosystems are units that are often arbitrary in their limits, but invaluable to science nonetheless. Human cultures consist of artifacts, such as knives of a certain shape and function; behaviors, such as initiation rites of a particular form; and mental constructions having little or no direct correspondence to reality, such as myths. The findings of anthropologists suggest that these units interact to form larger coherent wholes.

After a careful examination of this evidence, we realized that such elements form a natural base for the development of a theory of the coevolution of genes and culture. In order to increase the clarity of the unit, we matched it to the statistical definition of the "artifact type" commonly used in archaeology. The archaeologist, when presented with a mixed collection of swords or other relics, takes numerous measurements of their proportions, then compares the general appearance of all the specimens by a special statistical technique called multiple dimensional scaling. Those specimens close enough to each other to form a tight cluster are said to belong to the same artifact type. We decided to call the basic unit the *culturgen* (from the Latin *cultura*, culture, and *geno*, create) and to define it as a relatively homogeneous group of mental constructions or their products. In our classification the manufacture or use of a particular artifact type is therefore a culturgen. The same statistical procedure can be used to group other kinds of behavior and mental processes into culturgens. Psychological tests also exist that allow the comparison of these clusters with node-link structures in long-term memory. Overall, the culturgen can be used as a concept in both psychology and the social sciences.

Many culturgens are naturally distinct and would stand with or without theory. The preference for incest, for example, exists as a clear alternative to the preference for outbreeding. Women tend to carry infants on their left side close to the heart, a practice easily distinguished from other modes of infant transport. To raise the eyebrow in greeting is a gesture distinct from other facial signals. Additional examples occur conspicuously in most categories of thought and behavior. In other, less certain cases the statistical techniques of clustering can be used to make the results more nearly objective and open to confirmation. And finally, if the variation in behavior cannot be broken down easily in any other way, it can be described directly in the form of continuous mathematical functions.

These characterizations of the mental process are not scholastic exercises performed for their own sake. They are the necessary first step in the abstraction process—the most direct manner in which the largely visual picture of gene-culture coevolution can be turned into a series of precise models. The full theory can be converted into an explicit description with measurements and tested. As we began our own attempt, we did not expect everything to come out exactly right the first time, especially in the more complex cases of cognition and behavior. But we hoped to do better than previous scientists, who lacked any means of joining the evolution of genes and culture in a way that also takes into account the facts of psychology. Our aim was to find solid steps that lead from the world of psychology to the world of cultural anthropology.

CONSIDER THEN the view of cultural anthropologists as they enter the life of a primitive society. They are like naturalists stepping into the forest of a previously unexplored island. A flood of life envelops them, and they maneuver to establish the beginnings of some order and pattern within the framework of their own culture. The process can be seen in Charles Wagley's exemplary study of the Tapirapé Indians of central Brazil. At first there are the little details of quotidian existence:

> I took daily notes on men's economic activities, but found that a single man might do a dozen things in one day: he slept until 10 a.m., picked up his knife and axe and started toward his garden to clear underbrush; en route heard the call of a *jacu* (a black forest fowl, about the size of a chicken, which lives high in the branches of the virgin tropical forest) and spent two hours stalking it; arrived at his garden, worked for an hour, slept; returned to the village with one jacu; then spent the late afternoon weaving a basket in the takana.

As the exploration continues, Wagley becomes increasingly sensitive to personal relations and the nuances of language. The rainy and dry seasons pass by, and with them the rotating shifts in ritual and livelihood. Like a photographic image in developing fluid, the distinctive cultural traits of the tribe begin to emerge. The Tapirapé weep ritualistically when they greet visitors. They observe a strange order of meat taboos: women and children cannot eat wild duck, deer, armadillo, or jaguar, but everyone is welcome to the flesh of anteaters and cebus monkeys. There are men's societies named after birds, and spirits called anchungas who variously represent unborn children, the animals of the forest, and unnamed shadow spirits. The protagonist of their mythology is Thunder. He is a great supernatural power who leads demonic spirits called Topu in assaults on the Tapirapé people. In Thunder ceremonies shamans rise to meet the challenge and display their special knowledge and prowess. Wearing sunburst headdresses of scarlet macaw feathers, they and the novices smoke tobacco and enter trances in order to confront the unearthly adversaries. More often than not they are defeated—yet always with honor. Wagley's informant Panterí recounts a typical adventure:

> I ate much smoke and then smoked again. I sang; I saw one large sun. It came toward me and disappeared. I saw many small suns. They approached and left. I saw Thunder. It was small and came in a small canoe. It was Thunder's child (a topu). It wore a small headdress of red parrot feathers. It had a small lip plug. I reached to pull out the lip plug (thus he would have vanquished the topu) but it left the house. Then all was dark (he had not defeated the topu and it had consequently shot him with an arrow). I saw many suns. I travelled, singing as I walked . . . I saw Thunder; he is big and his body is covered with white hair. He had many feathers of the red parrot. I saw many topu and many souls of shamans near him. I did not talk but returned fast.

The anthropologist enters the dream world of the people; it is a special triumph for him to be initiated as a shaman or at least to be taught the priestly secrets. Piece by piece he constructs as complete a picture of the multilayered culture as possible. As the months pass he grows aware of yet more cadences and meanings, grasping many but never all the connections between the elements described in his notebooks. The entire culture often seems to fit together as an organic unit. The anthropologist therefore tends to speak of the "contextual" or "holistic" properties of culture and to stress the unique properties of each society in turn.

It would be enough to leave the ethnography—the straightforward description of the culture—at this point, laid out like a systematist's account of the natural history of a newly explored island. But more can be accomplished. The anthropologist can dissect elementary units out of the tissue of the description and treat them as culturgens. To take one example, the choice between incest and outbreeding can be selected for close study out of all the data on sexual practice. Because relevant information on the psychological development also happens to be available (from anthropological studies in other parts of the world), it is helpful to narrow the focus still more to brother-sister incest and its avoidance.

The essential datum with which to begin is the percentage of individuals who practice brother-sister incest over outbreeding, or at least prefer it if given a choice. The next step is the synthesis of field studies in order to obtain the percentage of incest choosers across many societies. Moving away from the Tapirapé and across the Amazon-Orinoco basin, the scientist encounters the Kayabi, Waurá, Kreen-Akarore, and Juruna, and still farther away the Jívaro, Siona-Secoya, Ye'kwana, and a score of other tribes. Although anthropologists seldom proceed in just this way, the combined cross-cultural information can be ideally combined into a single pattern, the *ethnographic curve*, which is what statisticians call a frequency distribution. It displays the percentage of societies with different degrees of preference. Thus a certain fraction of the villages has no members favoring brother-sister incest, another fraction of the villages has 1 percent of their members favoring this form of incest, still another has 2 percent favoring it, and so on. All these fractions together constitute the ethnographic curve.

The curve is a simple but precise measure of cultural diversity. How can we predict it from a knowledge of mental activity? In order to make the attempt let us imagine a process less easily studied than the mere statistics of choice: the thoughts of individual members of a preliterate society such as the Tapirapé. The knowledge on which they act moment by moment is stored as node-link structures (in other words, culturgens) in long-term memory. In the aggregate this information composes the culture. There is nothing mystical or intangible about the connection. However complex, however protean or ephemeral, a culture does not and cannot exist outside the physical structure of the brain and its artificial products. The following sexual reverie is based closely on data provided by Charles Wagley, but with the characters fictional and the actual train of thought speculative.

AT DUSK two close friends, Champukwi and Kamanaré, hang their hammocks side by side in the men's house. They notice with pleasure that they are alone, and lie back to relax, legs sprawling, eyes half closed. Their conversation runs lightly and in disconnected spurts over subjects that give the mind ease: of the place of spirits, Iungwera; of visits by Jaguar of the Skies; of jealousies and legendary battles against the dreaded Kayapó. The talk turns to sex. Champukwi's mind moves freely across the range of choices known to the Tapirapé. He remembers his personal preferences and adventures, matching them against the mores of the Tapirapé, and he guardedly reveals certain episodes and yearnings as he speaks to Kamanaré. He recalls the pleasures of extended foreplay with a woman, his body pressed and languidly moving against hers, his hands caressing her loins. He has heard of kissing from the white *torí*, but he laughs about that with Kamanaré: strange, disgusting! The same for cunnilingus: even worse! The *torí* may be joking. But fellatio is different; some do it, some don't, and it sounds pleasant. Almost any copulatory position is all right. Female orgasms sometimes occur, but who cares? They are the concern of the woman. Homosexuals are nice; they are treated as favorites on hunting trips, where they allow themselves to be used for anal intercourse. In the past a few dressed and acted as women and took a husband. That's all right—it's their business—but it *is* odd. Gang rape of women disowned by their families occurs from time to time. The other men join, but Champukwi won't admit to taking part. The whole thing is a little embarrassing. Adultery goes on all the time, and some do it a lot, especially Champukwi. But it creates complications; don't let your wife catch you, or the woman's brothers. It's something to boast about to a few other men, on occasions like this one. Champukwi has four lovers and has had many more in the past.

The haunting cry of a tinamou breaks the silence of the forest, and the conversation stops for a while. Champukwi thinks about the possibility of sex with his sister. Incest is forbidden implicitly. No one he knows has ever practiced it, and marriage between real brothers and sisters would be unthinkable. He himself does not consider his sister in the least attractive, although other men do. Champukwi imagines briefly what it would be like to copulate with her if they were to meet secretly in the forest or a distant garden. He winces. It could be done, he supposes, if other women weren't available and he had an intense sexual appetite, but it would bring little pleasure and a great deal of guilt. In this moment of decision, before the talk resumes, Champukwi has pondered two alternative behaviors and chosen outbreeding over incest.

IN COUNTLESS EPISODES the individual in each society reflects on competing culturgens. He imagines performing alternative acts, however fleeting and disturbing the images. He weighs their emotional rewards and guesses their impact on the minds of others. During all of these introspective periods a majority of individuals steadfastly reject incest and lean toward sexual activity outside the family. But a small minority occasionally desire incest. According to the stimulation and opportunities of the moment, they flip back and forth in their sexual preference. Occasionally, their incestuous impulses are translated into action.

Psychological studies on selected groups of people can ideally reveal not just the choice of individuals on a given day but also the magnitudes of their preferences. There are two culturgens available: preference versus rejection of brother-sister incest. Both reside in the long-term memory of the individual, to be called forth and pondered during reflection and talk. Given a preference for incest, what is the probability of a change to rejection? And given a preference for rejection, what is the possibility of change in the opposite direction, to a preference for incest? As the field anthropologist looks out over an entire village he envisages the population in flux: on a given day some members are holding fast to their choice between the two culturgens, while others are changing in one direction or another. And as the anthropologist considers many villages he is aware that these larger aggregates are also in flux. On Monday, for example, the fraction of villages with complete preference for outbreeding may change from 59 percent to 60 percent, while on Tuesday it shifts back again.

We reasoned that the probabilities for choice and the rates at which preferences change reflect the epigenetic rules guiding individual mental development. At the time of our study, information on epigenetic rules was limited mostly to the initial choices made by infants and children during the earliest bouts of learning and social conditioning. Some of these events influence later sexual preference and other forms of adult behavior. But such data can be misleading. The human mind is not insectlike, waiting to be imprinted on a single mode of action by early experience. The mind ruminates frequently on most or all of the choices stored in long-term memory. The most abhorrent options (the ones most opposed by the epigenetic rules) are turned away immediately, albeit with occasional feelings of quiet titillation. Those more nearly equal in appeal are played with at greater length. The mind flips back and forth between them before reaching a decision and taking the appropriate course of action. It also searches for new solutions and occasionally invents additional culturgens to be added to the repertory. Out of a vast number of such decisions across many categories of thought and behavior, culture grows and alters its form through time. The

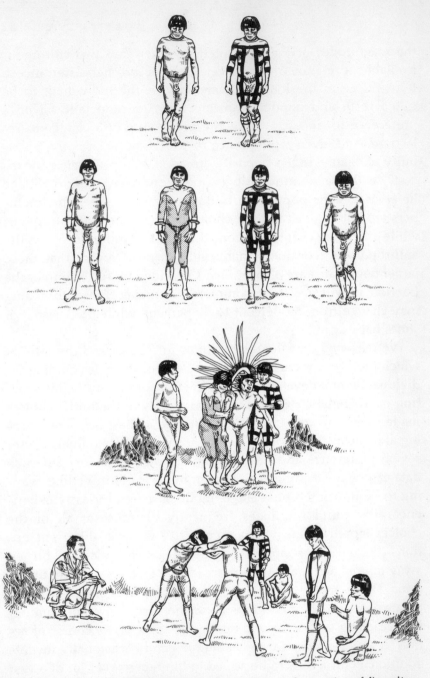

The stages that lead from individual decision making to the creation of diversity among cultures, as illustrated by body decoration in the Tapirapé Indians of Brazil. On the left are the activities as they would be observed directly by scientists. On the right the same processes are expressed in abstract form, in order to insert them into the quantitative theory of gene-culture coevolution. Proceeding from the top down, the sequence is as follows: the individual chooses whether or not to adorn his body, and he switches from one option to the other at a certain rate;

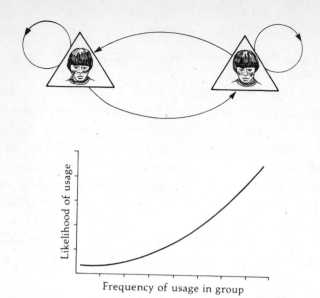

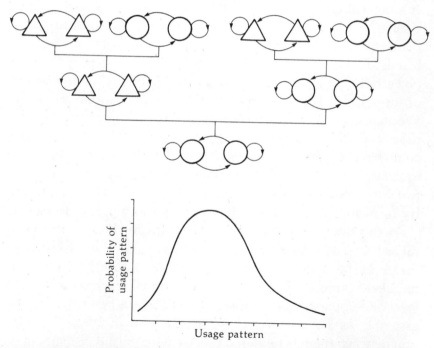

his rate of change depends on the frequency with which others express a preference for one choice or the other; each of the individuals in a kin group (illustrated in the third panel down) or society is either using body adornment or not; from the above information, the anthropologist (bottom panel) can estimate the probability that a certain percentage in the group uses adornment, that is, a particular usage pattern exists, at a given moment in time.

flux of culture comprises an unceasing torrent of changes in individual decision making.

The problem of translating mind to culture is similar to the challenge of part versus whole faced by scientists who study aggregations of subatomic particles, atoms, and molecules. A branch of theoretical physics, statistical mechanics, is devoted to the understanding of the holistic properties of these aggregations in terms of their component parts and the interactions among the parts. Mathematical techniques are available that allow remarkably accurate predictions of heat capacity, entropy, the condensation of gases, magnetic phenomena, and the diffusion of neutrons through matter. They have been responsible for many of the triumphs of modern physics.

It was tempting, at the initial stage of reconstruction of the mental process we and others had reached, to build models of cultural change by adapting the techniques of statistical mechanics. But of course mind and culture are more complicated than collections of inanimate matter. And what of history? Minds do not develop independently of other minds; they are powerfully influenced by the decisions already taken by the rest of the society. The resulting complications are subtle and deep. We had encountered a technical obstacle of imposing magnitude, and like climbers reaching the base of a sheer cliff, we could have turned back in good conscience. It is even fashionable to concede that history defeats theory. Many scholars feel comfortable with a compromise solution that accepts the rich tapestry of history as mankind's ultimate and imponderable nature.

All this doubt is understandable, but the methods of the natural sciences do provide a way up sheer cliffs. The pitons to be driven in the walls are the new assumptions added to the original, skeletal model. When the revised model reaches a sufficient level of complexity and realism, the climb can be resumed. The assumptions are then tested against the known facts. The relation between them is rendered precise with the aid of mathematical reasoning. Eventually the theory that spawned the model, in this case the theory of gene-culture coevolution, can be tested.

Our next step was to make the likelihood that an individual will switch from one culturgen to another (say from incest to outbreeding) dependent in some way on the surrounding culture. In order to describe that important complication, we developed equations related to those of statistical mechanics and included functions that reflect the sensitivity of individuals to the opinions and actions of others. Various assumptions about sensitivity were tried out, some obvious and apparently likely, others improbable. In one model individuals maintained a constant but low switching probability from the first culturgen to the second culturgen, until a certain percentage of the society came to favor the second culturgen; then the individuals jumped to a higher switching probability. In other models the trend watching among members of the society was made to be gradual—the likelihood with which a person shifted to the second culturgen rose steadily as more people adopted the culturgen. In still other models the reverse relationship was created—persons changed to a culturgen most readily when the fewest people around them accepted it: they rebelled. Or else they switched most rapidly when an intermediate number of their fellows favored the culturgen: they compromised.

Any of these imagined patterns of influence might exist in human societies. It seems probable that the precise form of the function—the level of peer activity at which the transition probabilities begin to change and the direction and rate of the change itself—varies from one category of thought and behavior to another. For example, attitudes about maternal care are likely to be more resistant to peer pressure and to change than attitudes about dress or cuisine. We searched the literature of social psychology to find out which patterns actually exist. This effort yielded very little information. Since no theory previously existed for which such data are crucial, psychologists by and large had not undertaken the necessary studies.

One exception was an experiment conducted on small groups of volunteers by Solomon Asch. He found that individuals do not like to be a minority of one, and they will often go against their own better judgment in order to conform to the general opinion of their peers. The likelihood that they will switch to the majority opinion rises steadily with an increase in the size of the group opposing them. Another social psychologist, Stanley Milgram, obtained similar results in an experiment on the drawing power of crowds. Under Milgram's direction, groups of volunteers met on a crowded street in New York City. On signal, anywhere from one to fifteen volunteers looked up at a tall building. The particular number of participants in each trial was arranged in advance. Separate observers counted the numbers of passersby who also turned their heads upward. The contagion of staring fitted the gradual-trend-watcher patterns: 4 percent looked up when they saw a single person do so, 16 percent followed five persons, 22 percent followed ten, and 40 percent followed fifteen. These and other fragmentary results could not be applied directly to the culture models we had created, but they gave a solid clue to the kind of functions that exist. They also illustrated the general form of small-group experiments needed to advance the whole subject.

We now possessed an array of mathematical functions, augmented by a very limited amount of evidence from social psychology, that led through one of the principal steps of coevolution—from the epigenetic rules of mental development to the formation of culture throughout the society. In principle at least, real information from psychology could be used to predict real information from anthropology, and vice versa. If psychologists monitoring the development of choice in a social setting were to produce the right kinds of data on switching rates and the effects of group pressure, they could tell the anthropologists what to expect in the statistical patterns of choice across many societies. And looked at from the opposite direction, the anthropologists might use such cultural data to infer the properties of individual mental development prior to psychological studies.

Were the psychology-to-culture models still too simple? If so, we could drive more pitons into the wall, add more complexity to the description of the basic mental processes. Even then, would we be on the right track? If not, greater complexity would yield only elaborate descriptions further and further from the truth. We needed to test the system by examining a real case. The requirements for this trial run were as follows. The mental processes and behavior should be simple and easily described. Data should exist on the transition probabilities between the alternatives available to individuals. The influence of other people and the surrounding culture on the transition probabilities must be measured. Finally, information should be available from many societies on the percentage of individuals choosing one culturgen as opposed to another.

All of these variables of cultural evolution are important enough to be studied on their own. In fact, when we looked into the technical literature of psychology and the social sciences, we found important reports on most of them. But the data were scattered in isolated clumps, and no one had pursued cause and effect from the beginnings of mental development to cultural pattern in a way that permitted the testing of quantitative models. It was the same old story, the catch-22 of scientific research: an absence of theory forestalls the building of theory.

One case, however, stood out as promising enough to develop further: brother-sister incest. We believed that most of the pieces were in place and that we might be able to approximate the missing ones. Recall the main features of brother-sister incest avoidance. The dominant epigenetic rule appears to be strong and clear-cut: when children are raised together in close domestic proximity during the first six years of life, they are automatically inhibited from full sexual activity at maturity. The rule is not absolute, but data from several societies, especially in Israel and Taiwan, suggest that it operates with at least 90 percent efficiency. We do not know the exact transition probabilities in sexual preference. But we do know from the Taiwanese studies that individuals practicing outbreeding are overwhelmingly inclined to continue it, while those practicing the psychological equivalent of incest (sexual relations with unrelated persons reared in the same household) are strongly inclined to switch away from it.

Thus we can safely assume a large difference between the two transition probabilities, with incest-to-outbreeding being much more likely than outbreeding-to-incest. Furthermore, these preferences proved to be relatively insensitive to social pressure.

The details in the Taiwanese case are the following. Before the Japanese occupation many families arranged "minor marriages" for their sons: unrelated infant girls were adopted for later marriage when the children reached maturity. In other words, the future couple were raised in close proximity, just like brother and sister, even though it was clearly understood from the start that they were to become husband and wife. The arrangement worked out poorly in comparison with traditional marriages between other couples in the same villages. Divorces and infidelity were more common, and fewer children were produced. Generally the couples linked in minor marriages strongly preferred sex with partners of their own choice over the "incestuous" relationship, even though the society around them approved the practice and their relatives often tried to force them to consummate the marriage. Such insensitivity to the preferences of others probably occurs in very few other categories of human thought and behavior. But it was a perfectly respectable extreme example with which to begin. It greatly simplified the analysis and permitted us to devise the first estimates for a real case of translation from individual mental development to cultural pattern.

The results are shown in the accompanying figure. The curves displayed are the ethnographic distributions—the percentages of societies expected to show various degrees of brother-sister incest or at least preference or acceptance of it. Some of the findings illustrated in the figure are of more than ordinary interest. First, it is an important working principle of the natural sciences that a properly constructed theory allows for all reasonable possibilities and does not argue exclusively for one final result. By adjusting the likelihood of changing back and forth between incest and outbreeding, different ethnographic curves are obtained. The test of the theory comes from the fact that, once these probabilities are learned from direct observations (in this case

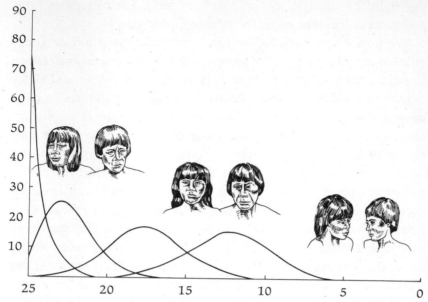

The diversity of brother-sister incest across many cultures, as predicted by the theory of gene-culture coevolution. Passing to the right, most persons prefer brother-sister incest; to the left, most prefer to mate with others rather than their siblings.

from studies of child development), the result at the next higher level (in this case culture) is specified and can be compared with the facts.

In considering all such possible worlds, we can start by imagining a human population in which individuals are inherently indifferent in their choice of partners. Brothers and sisters are considered just as sexually attractive as outsiders. The expected curve is the one shown in the center of the figure, with a bias value of 0.5. In other words, 50 percent of the time a typical individual preferring incest will shift to outbreeding and 50 percent of the time he will continue to prefer incest. In other imagined populations, the genetically based epigenetic rules turn individuals against incest. The transition probabilities may be moderately in favor of outbreeding (0.7 as shown in the second curve from the right) or strongly in favor (0.9 or 0.99). Each of these rates produces a different expected ethnographic curve.

We now come back to the fundamental conception of gene-culture coevolution and the underlying force of natural selection. Breeding between people as closely related as brother and sister results in a more frequent occurrence of genetic defects in children and a correspondingly lowered reproductive rate. Studies conducted by geneticists working independently in several countries have shown that the damage is usually significant and often catastrophic. As a direct result, natural selection is expected to operate so as to reduce the occurrence of brother-sister incest. The transition rate moving individuals away from incest (the number displayed in the figure) is the factor changed over many generations by this natural selection. If the selection is very mild, the result expected when genetic defects are rare, the transition rate might remain neutral or evolve so as to favor outbreeding to a moderate degree. If the selection is severe, the transition probability from incest to outbreeding can be expected to evolve to a higher level, say 0.9 or above. The result will be cultures in which brother-sister incest is relatively rare.

Where in fact does the human species fit in this array of theoretical possibilities? The developmental data suggest transition probabilities greater than 0.9. We can now look for information on the amount of brother-sister incest in various cultures, in order to see whether the translation model works. Although few exact data are available from studies in cultural anthropology, anecdotal accounts are available from societies as diverse as Tikopia and New Guinea in the South Pacific, Ghana in West Africa, Israel, and the United States. They suggest that the true enthnographic curve is closer to that predicted by an assumption of strong developmental bias. In the figure it would be nearer to the theoretical curve generated by a transition probability of 0.99 (away from incest and toward outbreeding) than to the three others, which range from 0.5 to 0.9. So there is at least a crude match between fact and theory in the one case where the theory can be put to a test. New information gathered in the future will allow the theory to be tested with increasing severity.

An effect of equal importance revealed by the theory is the creation of diversity among cultures even when the properties of individual mental development are genetically determined. Examine the curve for an anti-incest bias of 0.99, which is believed to be in the range of that actually existing in human beings. Suppose for argument that this key parameter is fixed at precisely that figure out to many decimal points (0.99000 . . .) and in every person around the world. In other words, the brain of each individual is constructed to operate at exactly this transitional level and no other. It might seem to follow that societies will also be completely uniform in their composition, with 99 percent of the members against brother-sister incest and 1 percent favoring it. But this would not be the case, since a substantial amount of cultural diversity will be created by chance alone. The various cultures will drift back and forth. In an array of societies each containing twenty-five members about 80 percent of the societies would be wholly free of any pro-incest feelings, while the remaining 20 percent would contain one or two individuals favoring incest.

If the population of each society is larger, the number of such groups whose members are uniformly against incest will drop still further. Among groups of seventy-five persons, only 46 percent will be free of incest. When the bias is less, the diversity increases. Suppose that the human species were genetically fixed to be perfectly indifferent in sexual choice: the probability of transition is exactly 0.5000 . . . , again to many decimal places. The result would not be 50 percent of the people favoring incest and 50 percent against it across all societies. Instead, there would be an impressively wide range of diversity, as shown in the figure.

This statistical scattering will come as no surprise to mathematicians who deal with similar operations daily, but it does run counter to the intuition of many others who work professionally on human behavior. Whether it is construed correctly is of major importance in social theory. Scholars often argue that the very existence of cultural diversity demonstrates that biology cannot play a guiding role in human social behavior. But it is clear that cultural diversity can be generated in the presence of even rigid genetic controls. What is determined is not a particular social response but rather the statistical *pattern* of response across many societies. Only quantitative models incorporating a knowledge of development can predict this pattern, and exact ethnographic information gathered in anthropological field studies is needed to test the models. The symbiotic relation between psychology and anthropology is thus established. The pattern of cultural diversity should not be employed to deny biological influence, but to illuminate it.

We can now ask what happens when history plays a more commanding role than in the special case of incest. Societies respond to the perceived demands of different physical environments, and they are affected profoundly by the cultures passed on to them by previous generations. Their reactions to these environmental influences may increase the amount of cultural differences among societies beyond that expected from chance alone. But it is also easy to imagine sets of circumstances in which cultural diversity will be reduced.

Again the problem can be attacked in the stepwise manner of the natural sciences. The models used to solve it can be made gradually more complex. In biology there is a useful strategy of research known as Krogh's rule, after the great Danish physiologist August Krogh: for every problem there exists an organism ideally suited for its solution. *Drosophila* fruit flies have proved ideal for genetics owing to the ease with which they can be cultured and the shortness of their generation time; squid have been ideal for nerve physiology because they have giant nerve cells that can be readily isolated and studied. The principle can be modified for the analysis of culture: for every new, more complex stage in theory construction there exists a category of culture ideally suited to test it. In moving beyond incest, the investigator should search for cases that are somewhat—but not excessively—more complex.

The logical next stage in our thinking was to stay with only two cultural choices but to make the decisions of individuals dependent on the preexisting cultures. In the models described earlier, the transition probabilities from one cultural choice to another changed according to the percentage of other members of the society who favored one choice over another. Consequently the rate and direction of change in the society depend in a sensitive manner on the previously existing culture. Other elements entailing history, such as population density and food resources, can be added one at a time as required by the pattern of the culture under analysis. Finally, cultural innovation can be incorporated into the models in the form of new cultural choices that compete with those already in existence.

To state this approach as concisely as possible, we conceive of the study of history as ideally consisting of something more than the description of a large number of unique events, more than a generalization of the causes and effects of broad changes in culture. It must also consist of the breaking apart of cultural change into elements and their analysis by deductive models. The analytical phase will be by far the most difficult in future historical studies. The trick is to classify the elements and identify the correct sequence in which to link them.

LOOKING FOR the best examples of social behavior, the equivalent of Krogh's ideal organisms, we continued to direct our pilot studies to more primitive societies. It seemed to us that an especially interesting and favorable historical process might be found in the multiplication of villages by the Yanomamö Indians, as described by the anthropologist Napoleon Chagnon. The approximately 15,000 Yanomamö live in the forest areas of southern Venezuela and adjacent portions of Brazil. They are organized into villages containing from 40 to 250 people, separated by a distance of about a day's walk. The members of each village are tightly linked to one another by intricate bonds and rituals of kinship, and all are in close daily contact.

The Yanomamö have been called the most aggressive people on earth. Frequent bloody raids between villages account for as many as one third of the deaths of adult males. These wars are almost always over women, since the acquisition of wives by barter, raiding, and seduction is a goal inflated by an unusual emphasis on polygamy. The ultimate reward of social success, in both a psychological and biological sense, is to own many wives and to father many children. As a rule, headmen have about twice as many children as other men in the same village. When Chagnon asked the Yanomamö why they fought, this answer was typical: "Don't ask me such stupid questions! Women! Women! Women! That's what started it! We fought over women!" The frequent raiding is accompanied by a high level of squabbling within the villages, often over women but also involving both sexes in disagreements over status, food, and various domestic issues.

For at least a hundred years the Yanomamö population has been growing in size and expanding its range. The opportunity to expand appears to have come to these feisty people through the destruction of surrounding tribal groups by European civilization. The Yanomamö can be compared to primitive people invading a previously uninhabited land. They also serve as a model of early humanity in their economic organization, since only within the last few centuries have they shifted partway from a hunter-gatherer subsistence to an elementary form of slash-and-burn agriculture.

As the local population grows, villages split, move apart, and occupy an overall larger territory. The split occurs when one group, usually consisting of one or more extended families, leaves the mother village and moves to a new site. The fission is the result of accumulated strife and tension, which occur in all the villages but grow in frequency and intensity as the population gets larger. A point is reached when a Yanomamö village can no longer be held together by the bonds of kinship and the relatively weak authority of the headman. A group of kinsmen then departs in resentment and fear. A new village is built, leading to new tribal lineages, wars, alliances, counter-alliances, and acts of treachery. In 1970 Chagnon, speaking in the Yanomamö tongue, discussed matters with Dedeheiwä, a member of the Mishimi-shimaböwei-teri:

"Did your ancestors split away from the Karawatari?"

"They split away, split away, split away. My own ancestors split away in the past and left the Karawatari."

"Where—what garden—did this occur?"

"Over in that area, at Shihenaishiba garden." Dedeheiwä points toward the forest.

"Did your ancestors and the Karawatari ancestors live together at Shihenaishiba garden?"

"Yes! Yes! Yes! Yes! Yes! My true ancestors were attacked mercilessly at Kayurewä garden by the Aramamisi-teri. Then my ancestors, the ones attacked, fled in fear away from that place and moved right past the very spot where this garden [Mishimishimaböwei] now lies."

"Are the Aramamisi-teri and Kohoroshitari closely related to each other?"

"Yes! Yes! Yes! Yes!" Dedeheiwä turns and gestures toward the Shanishani River. "The Aramamisi-teri chased them away from there."

"Where did their war first begin?"

"Way, way up there, in the very headwaters of the Shukumöna River . . . past the spot where the headwaters of the Rahuawä River come close to the branches of the Shukumöna."

"Did they all once live together way up in that region?"

"Yes . . . then they fought with each other with arrows and had a war, and then the Kohoroshitari fled."

With an increase in tensions, violence in the Yanomamö villages develops through three levels, each requiring reflection, calculation, and moment-by-moment decisions. The least dangerous form of fighting is the chest-pounding duel, in which men trade blows with their closed fists. The combatants challenge one another in response to malicious gossip, stinginess in trade, or almost any other petty grievance. Anger abates, at the cost of bruises and blood that may be coughed up for days afterward. The next level of aggression is club fighting, which is roughly equivalent to chest pounding but with a weapon. The wooden clubs are about the same shape as a pool cue but twice as large, six to nine feet in length. Most duels begin when one man catches his wife having sex with another man. The enraged husband challenges the rival to strike him on the head with a club. He stands still, holding his own club vertically. After taking the blow, he delivers one of his own. At this point, as blood streams down the face and neck, the fighting often becomes wilder, and the men try to strike one another on other parts of the body. Others join in, usually taking sides according to kinship but sometimes just to restore a fair balance. In late years the cranial dents will be proudly exhibited as evidences of bravery and manhood, as the cheek scars of Heidelberg saber duelists used to be.

In the electric atmosphere of frequent chest pounding and club fighting, a single incident can trigger the decision to split the village. One club fight, culminating a long series of incidents, will be enough. Or the community may hold together long enough to witness a still more violent episode, at the third level of aggression. Such was the case of Patanowä-teri, as witnessed by Chagnon.

It began when one young man stole the wife of another, who was allegedly mistreating her. The two rivals then engaged in a brutal club fight. Tensions had been running high in the village, and soon almost every other man had joined the struggle. The headman of Patanowä-teri tried to control the melee by restricting the fighting to clubs. But the young man suddenly and unexpectedly speared and wounded the husband. Other men began to jab at each other with the sharpened end of their clubs. Enraged at the outcome, the headman ran his club through the body of the young man, who died when others tried to remove it. The wife was given back to her husband, who punished her by cutting off her ears with his machete.

The relatives of the dead man were then ordered to leave Patanowä-teri to avoid further bloodshed. They traveled to the villages of the Monou-teri and Bisaasi-teri, natural enemies of the Patanowä-teri, who promised to give them temporary protection in exchange for several women. From there the refugees plotted with their hosts to conduct raids against the mother village, in order to get revenge and kidnap other women.

The critical psychological level causing division of a Yanomamö group comes most frequently when the population exceeds a hundred individuals. Fissioning is rare in villages with fewer than eighty members, regardless of the amount of internal turmoil. The reason is that a village must have at least ten able-bodied men to engage in raiding and defense. If the raiding party is smaller, it is likely to lose pitched battles. In ordinary circumstances this large a male group exists only when the overall population of a village is about forty to sixty. So a village must contain at least twice that number, or approximately one hundred members, in order to divide into two self-sufficient units.

Historical circumstance is a prominent factor in village fissioning. The present-day Yanomamö started in a South American rain forest, utilizing a relatively productive mode of slash-and-burn agriculture, organized into scattered villages only somewhat more complicated than bands of hunter-gatherers, and with little competition from other Indian tribes and Europeans. All this freed them to expand their populations, at least for a few generations. The men have accordingly extended the practice of taking several wives, which is the prevailing practice of all hunter-gatherer and primitively agricultural people, into an extreme form that promotes frequent outbreaks of warfare. Variations on all of these ethnographic themes are to be found in other economically primitive people around the world, but many details imposed by the environment are peculiar to the Yanomamö.

There is history too in the details of the Yanomamö expansion. The Yanomamö live in villages bearing such names as Patanowä-teri, Reyaboböwei-teri, and Iwahikoroba-teri (the suffix -teri is similar to *pueblo* and can mean either village or people). These communities may seem very small and insular to us, but for the people inhabiting them they are complete worlds, the theaters of a rich and exciting existence. The villages were created by a particular succession of overland treks, wars, outbreaks of disease, heroes, notable marriages, food bonanzas, and legendary feats, upon which the Yanomamö mind dwells with complete absorption. The history of Patanowä-teri is different in scale but not in kind from the history of England, and Chagnon is its Macaulay.

True to the Krogh principle, the difference in scale makes it possible to analyze the history of the Yanomamö villages deductively so as to reveal something of the relation between individual behavior and culture. As a first step, fissioning can be treated as a binary decision made by individuals: to stay together or to separate. The likelihood that one of these options will be chosen over the other is affected by both the opinions of others and the size of the village population. During conversations and disputes both aspects are reviewed repeatedly. Individual Yanomamö change more easily from "depart" to "stay" when the village contains about one hundred people or less, giving way to the strong reverse tendency when the population size approaches two hundred and protracted violence flares. Chagnon observed that villages steadily expand in size until they reach an apparent "critical mass." When the group is small, squabbles die down quickly and individuals are relatively unresponsive to isolated instances of strife. But above the critical population size, small confrontations inflame and spread rapidly throughout the village. When enough people prefer to depart, one family group or another leaves to set up camp elsewhere, which in effect divides the community into two villages.

From the accounts of the fissioning process by Chagnon, we constructed a model of Yanomamö thought that appears to capture the essential details of individual decision making. It goes beyond the incest model by incorporating the effects of peer opinion and the rising tide of discord caused by population pressure. The key property is this: at some level of behavior on the part of other members of the tribe, the probability that an individual will change his attitude abruptly rises or falls. When village size is small, the probability that a village member will change from "stay" to "depart" remains slight unless an overwhelming majority of the group favors splitting up the village. After this majority is reached, decisions to depart are made somewhat more readily, but there is still considerable reluctance on the part of all members to disrupt the social structure. Thus groups in small villages are strongly prone to arrive at a decision **to stay, even** after they have made an initial decision to depart.

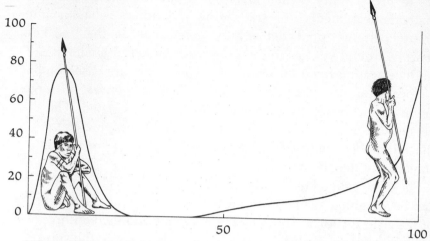

Village splitting in the Yanomamö of South America, as expressed quantitatively from the theory of gene-culture coevolution. Passing to the right, most individuals prefer to divide the village; to the left, most prefer to stay and keep the village together.

They are still more likely to revert once the number of stayers has reached a small minority. They are easily persuaded to keep the village intact. Hence the ethnographic curve for a small village is peaked sharply around patterns of group decision in which the number of members wishing to depart is small.

This is not the case when the village is large. The internecine squabbling lowers the flash point at which individuals are more likely to depart. The predisposition increases overall as conflict and violence intensify. The number of departers required to increase this predisposition to near certainty becomes smaller and smaller. In a strife-torn village, the sight of even a few peers preparing to leave is enough to resolve one's own will to do so. The likelihood that *n* departers will change their preference and remain in the village decreases as the village grows in size. In order to raise the probability of one's changing from depart to stay even by a small amount under these conditions, it would be necessary for a very large proportion of the remainder of the villagers to prefer unity. When given such psychological rules, our mathematical theory for translating individual behavior into group patterns predicted ethnographic curves that peaked sharply around villages in the process of violent fission.

Once again, even if the changes in village size are based on biologically fixed schedules of individual responses, a large amount of cultural diversity will inevitably be created. The diversity can be measured in terms of the ethnographic curve, or the relative frequency of societies with various percentages of stayers and departers.

For more general use, this kind of anthropological information can be weighted against data on the psychological development of the crucial choice. Then the theoreticians and field anthropologists can work as a team to probe more deeply into the subject. More complex models can be perfected to correct and extend the theory. New factors, such as the abundance of food and the power of headmen, can be incorporated into the analysis and their influence weighed. As the study becomes more intricate and incorporates finer details and unique events, anthropology will blend into more conventional aspects of the study of history.

The Yanomamö case illustrates the considerable promise of the analytic approach contained in the theory of gene-culture coevolution. But it also reveals starkly the inadequacy of most anthropological data for quantitative theory. The decision rules defined in village fissioning are almost certainly a compound of separate epigenetic processes of mental development that variously entail aggression, xenophobia, and bonding to relatives. These processes need to be teased apart by means of close studies of the socialization and maturation of attitude in children and young adults. The ethnographic curves expressing the diversity of opinions across the Yanomamö villages can only be guessed at right now. The surveys needed to make them precise have yet to be conducted.

A concern is sometimes expressed among anthropologists that their science has peaked and begun an unavoidable decline.

Most of the world's surviving primitive societies have already been studied, and libraries are filled with the greater part of the information worth recording. But if gene-culture coevolution is explored, if the influence of the epigenetic rules on culture can be measured, and if biology therefore has validity for the human sciences, anthropology is destined to enjoy a powerful resurgence. It is in every respect an infant branch of the natural sciences. The facts already gathered will remain valuable, but vastly more are needed to construct a full explanation of the linkage between biological and cultural evolution. As that new form of explanation matures, it might be possible to develop a comparable analysis of many aspects of the history of more complex societies, up to and including modern industrial states. Yet because the matter is so easily misconstrued, we cannot stress enough that the existing theory and methods are not adequate to treat more complicated patterns of behavior and the institutions of advanced societies. Human sociobiology is in approximately the same position as molecular biology in its earliest days. That is, several key mechanisms have been identified, enough to explain elementary phenomena in a new and more precise way. The subject is still rudimentary, but if both biology and culture are to be taken into account, it seems the only way to go.

## The Second Challenge: Culture to Genes

With care we can proceed from biology into culture in a few categories of behavior in simpler societies. But then comes the much more difficult task of tracing culture back into biology. How is the evolution of the genes affected by the cultural environment in which they are embedded? The possible nature of this interaction can be visualized in concrete form by traveling back in the imagination 25,000 years, to a time when the human revolution was in full progress. Art, counting, religion, and perhaps language itself were being invented. Man had become fully *Homo sapiens* and was spreading out of Africa and Europe to the far reaches of Earth. Soon bands would cross the Bering land bridge from Siberia to Alaska and colonize the New World. Others would press southward into New Guinea and Australia.

During this critical period the mind continued to evolve new epigenetic rules. In each new form of endeavor, individuals needed to learn the best cultural choices and to learn them quickly and well. Important qualities such as deep grammatical structure and the emotional response to tribal ceremonies were not left to the vagaries of culture and parental guidance. They and other vital mental functions were enciphered and nourished by genetic programs in the brain. As mental life became more complex, so did the algorithms and constraints that conditioned it.

The hunter-gatherer bands perfected their stone knives and spearheads. They carved ivory splinters into fish hooks and sewing needles. They invented names for food plants and dreamed the mythic stories of animal spirits to be told around the campfire. And as the culture proliferated, so did genetic diversity within local populations of the species. Even as the rules of mental development were growing in strictness and power, societies as a whole were moving toward balance and flexibility. There is a theory for this seeming paradox. A few cognitive traits are so superior that the epigenetic rules producing them will evolve so as to be highly developed in everyone. Almost every person, for example, can be expected to acquire language at a speed that would seem miraculous to a chimpanzee. People invent words at a prodigious rate to label sights and sounds (but not odors). With near unanimity they enjoy sugar, avoid excrement, and marry outsiders.

But in other cases there is a limit to the benefits of particular traits: more is not necessarily better. Many circumstances exist in human life in which moderation is beneficial and excess destructive. Aggression is a prime example. Too little on the part of individuals encourages domination by more aggressive peers and the weakening of overall tribal resolve; too much causes recklessness and self-destruction. The resulting compromise can lead not only to intermediate levels of hostility but to a greater diversity among people in the way it is handled.

A wrestling match in a Yanomamö village.

Most present-day groups of primitive men engage in some form of organized aggression, and it is reasonable to suppose that ritualized fighting and war were also widely practiced by our ancestors 25,000 years ago. The forces that lead to moderation and diversity can be seen at work in the Yanomamö, as revealed by a typical incident in the history of the Monou-teri.

THINGS BEGAN to go downhill when the Monou-teri men attended a celebration by the Patanowä-teri and Bisaasi-teri to which they had not been invited. Upon arriving they discovered seven unescorted Patanowä-teri women outside the main village. The temptation was too great: they seized the women and carried them back to their own village. The next morning the Patanowä-teri men went to Monou-teri armed with clubs. In the course of the ensuing fight they recovered five of their women. But they could not get the other two without shooting arrows to kill, and the headman was unwilling to escalate the conflict to this final stage. He decided to hold the aggression to the second level, club fighting, because the Patanowä-teri already had as many enemies as they could handle.

The Monou-teri of course knew of the need of the Patanowä-teri to compromise, and they had calibrated their own response accordingly. But there was one factor that threw their calculations off. The headman of Monou-teri was Damowä, a man of especially violent temper and aggressiveness. He was the only true *waiteri* (fierce one) in his village. Damowä, enraged because the Patanowä-teri had wrested back so many of their women, led a full-scale raid against them. The war party caught a Patanowä-teri warrior, Bosibrei, near the village, as he was climbing a palm tree to gather fruit. The raiders killed Bosibrei with a single volley of arrows, then ran for home.

Now the Monou-teri prepared for inevitable raids by the vengeful Patanowä-teri. Keeping constant watch, they began to clear new gardens across the Mavaca River to serve as retreats. But fierce Damowä was careless and while on a honey-collecting expedition got trapped by a Patanowä-teri party. In the ensuing fight he was killed, shot full of arrows.

The deaths of Bosibrei and Damowä changed Yanomamö history. They triggered a series of new political events that harmed both of the warring groups. The Monou-teri had difficulty finding a new headman and finally settled on Orusiwä by default. As a result leadership was shifted from one lineage to another. Both the Monou-teri and Patanowä-teri frequently moved their villages in an effort to defend themselves. As the raids continued and embroiled still other tribes, the fiercest men, the *waiteri*, were steadily killed off. Seen from the vantage point of evolutionary biology, the conditions were ideal for the genetic evolution of moderated aggression and a diversity of mental devices for managing conflict.

SYSTEMATIC COMPROMISE is favored in other activities besides aggression. In hunter-gatherer societies such as the !Kung of the Kalahari, conspicuous attempts to improve personal status and to accumulate large quantities of personal goods are met with ridicule and hostility. The result is the maintenance of a nearly egalitarian social order. In economically more complex societies, specialization and division of labor are responsible for another trend in moderation. Excessive production of goods and services of any one kind leads to an increase in competition, a destabilization of the market, and ultimately a reduction in net benefits to the specialized producers. Rising costs of processing and transport can also precipitate a decline. The ultimate result over a sufficient number of generations will not only be a spreading of economic and social roles, as expected from elementary economic theory. It will also include a diversification of the genes that create the capacity to assume each role separately.

This does not mean that human societies are evolving into a genetic caste system. The powerful tendency to select mates outside the immediate family and to exchange young people between tribes militates against the full development of such a trend. The likelihood of partitioning is further reduced in more advanced societies by social mobility and opportunities to change occupations. Even the caste system of India, which is the most rigid and elaborate on Earth and has persisted for two thousand years, is maintained largely by cultural conventions. So far as is known (although the matter has never been thoroughly studied), members of different castes differ from one another only slightly in blood type and other measurable anatomical and physiological traits. What the analysis does suggest is that a greater diversity can readily develop among people in the developmental rules that influence the selection of social and economic roles. These propensities, and the genes underwriting them, exist in a state of flux in populations. Although they cannot easily carve up societies into biological subgroups, they do maintain a higher level of genetic individuality. Natural selection appears to have been a prolific and generous creator, producing many talents and lavishing them across the population. If that much is true, it follows that each person is even more distinctly an individual than scientists and humanists alike had previously guessed.

In the evolutionary crucible of 25,000 years ago, genetic changes in the brain could have occurred one upon the other in rapid succession. The coevolutionary equations suggest that a "thousand-year rule" was operating, which can be expressed roughly as follows. In as few as fifty generations—about a thousand years—substantial genetic evolution can occur in the epigenetic rules guiding thought and behavior. To see more clearly what might have happened, let us go back in time a few thousand years to a period of rapid evolution. Suppose that at the beginning of the episode a certain epigenetic rule was strongly influenced by one gene and another epigenetic rule by a second, competing gene. The first epigenetic rule predisposed individuals to choose one culturgen—say flaring of the nostrils, to denote displeasure—and the second epigenetic rule predisposed them to choose an alternative culturgen, say frowning, to indicate the same emotion. In the example at hand the population started with a preponderance of the first gene, the first epigenetic rule, and a tendency to flare the nostrils. If frowning during displeasure conferred even a small survival advantage on those who use it over nostril-flarers, roughly a thousand years would have been needed to substitute the second gene for the first. At the end of that time, fifty generations down the line, the population would consist mostly of people who possess the second gene, epigenetic rule, and facial expression. So—in theory at least—the past 25,000 or 50,000 years might well have seen important evolutionary changes in the human mind. We don't know whether in fact this did happen, but it should now be considered at least an excellent possibility.

The overall genetic change must have been greatly accelerated by cultural innovation. The feedback from culture to the genes is the one evolutionary process probably unique to human beings. In most of the millions of noncultural animals, behavior is prescribed unilaterally by the genes. The patterns of behavior are rigid and under direct genetic control. Each species has a peculiar repertory of such "instinctive" behaviors, and fidelity to the repertory closely adapts members of the species to a particular niche in the environment. Thus the house fly searches through open space for the odor of decaying fruit of certain kinds; it lowers its proboscis in just a certain way to feed on sugars; it mates in response to a narrow set of courtship signals; and it lays eggs in sites characterized by certain odors and levels of humidity. The lockstep nature of its responses allows the fly to thread its way brilliantly through an enormously complicated and dangerous environment.

Evolution in insects and other small-brained animals occurs when spontaneous gene mutations or recombinations of previously existing genes produce a novel pattern of behavior. If this altered repertory imparts higher survival or reproductive ability to the animals that display it, the genes prescribing the behavior will spread through the population. By definition, the species then evolves from the old repertory to the new one.

Human evolution is both similar to and profoundly different from that of the fly. Although scientists and philosophers have generally understood that culture is a new force in the evolution of life, its capacity to transform heredity is far greater than has been generally appreciated. The conventional picture of human evolution is one in which culture *replaced* genetic evolution: the brain evolved to a high level of capacity in memory and reasoning, all of a relatively generalized form, whereupon culture took over and guided mankind the rest of the way. According to this view, the past 25,000 years of history has consisted of dead reckoning with hereditary equipment acquired in earlier ages. But the evidence from cognitive psychology and the models of gene-culture coevolution point to a radically different relation between genes and mind. Although new forms of behavior, the "mutations" of culture, are invented by the mind, which forms of innovation occur is very much influenced by the genes. A feedback then occurs from culture to the genes. The greater biological success of certain kinds of behavior causes the underlying epigenetic rules and their prescribing genes to spread through the population. Genetic evolution proceeds in this manner, rendering future generations more likely to develop the particular forms of thought and behavior that imparted success during the earlier tests by natural selection.

Gene-culture coevolution is a radical process. The differences among individuals made possible by higher intelligence and culture enormously increase the potential for natural selection and genetic evolution. No longer do the genes dictate one or a very few behaviors—instead the mind intervenes decisively. Ranging widely, it creates a much greater array of actions. It permits each combination of genes to have multiple expressions and offers alternative solutions to most problems within a single lifetime. Because the mind judges the circumstances of the moment and reflects on the broader implications of its own decisions, it can address the ordinary contingencies of life with greater moment-to-moment competence than the tightly programed responses of animals. In effect, the mutation rate of behavioral responses is vastly increased, and the genes that capacitated the responses are tested more rapidly than in ordinary, instinct-guided species.

As a result, evolution is accelerated. The genes continue to hold culture on a leash; in each generation the prevailing epigenetic rules of mental development affect which cultural innovations will be invented and which will be adopted. Yet culture is not just a passive entity. It is a force so powerful in its own right that it drags the genes along. Working as a rapid mutator, it throws new variations into the teeth of natural selection and changes the epigenetic rules across generations.

### The Third Challenge: Why Only Man?

If correct, this blend of theory and factual evidence has brought us closer to an understanding of what is truly distinctive about human change and why it has been so quick. But there remains the nagging question of why man alone took the fourth great step of organic evolution. The fossil beds are dotted with the remains of large-brained animals that might have achieved the same thing earlier. One hundred million years ago, fifty times farther back in time than the appearance of the earliest true men comprising *Homo habilis*, large ammonites and other archaic relatives of the squid and octopus swam the Jurassic seas. Their large saucer-shaped eyes surveyed the water around them, and their tentacles played over the coralline and mud surfaces of the ocean floor. What might they have been thinking? Perhaps there was a mind of sorts, and their brains worked to enlarge and exploit the limited amount of information already stored in their associative nerve cells. On the land lived human-sized dinosaurs who walked semierect on their hind legs. They possessed relatively large brains and might have manipulated objects in their three-fingered hands. Surely they were prime candidates for the ascension to high intelligence and culture. "Dinosauroids," as the paleontologist Dale Russell has called their imaginary brilliant descendants, could have beaten man to the tape by a hundred million years, but the opportunity passed. The great cephalopods and reptiles became extinct, and large-brained mammals proliferated in their place. Ten times farther back than the origin of man, the African savanna on which that unique event was to occur swarmed with numerous elephantlike forms, hyenas, monkeys, and apes. None managed to enter the self-propelling circuit of gene-culture coevolution. Millions of species passing through hundreds of millions of generations comprised of uncountable billions of individuals, faced by every conceivable environmental challenge and opportunity, shuffling astronomical numbers of genes in microevolutionary experimentation—all this immense ferment managed to push exactly one species across the threshold and into the autocatalytic climb to advanced culture. Something very peculiar and powerful must have been holding the evolving systems back.

The great anti-culture obstacle has something to do with the sheer mass of information that has to be stored in a single brain to achieve advanced learning. For each generation embarking upon it, gene-culture coevolution is a daring series of leaps into the unknown. Knowledge has to be renewed and behavior

An evolutionary advance that did not occur. On the left is a reconstruction of the bipedal dinosaur *Stenorhynchosaurus* which lived near the end of the Mesozoic Era and had some of the traits thought to make the origin of advanced intelligence possible. On the right is the "dinosauroid" as conceived by Dale Russell. This creature might have evolved from *Stenorhynchosaurus* a hundred million years before man—but did not.

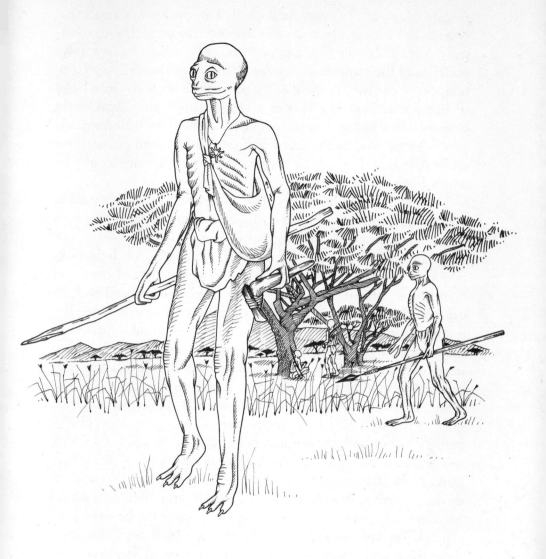

freshly developed in every individual. There is always a risk that enough knowledge can be lost or distorted to break the continuity and endanger the entire population. We have seen how local societies of the transitional species *Homo erectus* actually suffered temporary declines in the quality of their stone tools. In a band of thirty members, the loss of a single gifted individual could have reversed cultural evolution and doomed the little community to extinction. At the dawn of intelligence, when populations were sparse and life most fragile, an entire species could be snuffed out.

There is also a purely physiological risk to individuals having a more complicated brain. At first glance it might seem that evolution needs to create nothing more than a big brain—one with extra association areas to enlarge the capacity for long-term memory and basic reasoning. Perhaps the structure of the brain could be simplified somewhat, as instincts are discarded and the organism becomes a generalized processor of information. But not at all—the exact opposite is the case. The mind deprived of the automatic guides of instinct still cannot choose blindly. It can never survive as a blank slate. It has to make the correct choice at each crucial moment, and to do so it depends on a whole new array of machinery made out of nerve cells. These mechanisms guide it through the correct procedures of memory, concept formation, valuation, and decision making.

The distinction between the two basic strategies of mental action can be illustrated by the manner in which smells are handled by animals. Insects mostly respond in an automatic fashion to odors. The male is attracted to the female by her sex substances, the female lays eggs on the leaf of a food plant that emanates the correct terpenoid smells, and so forth. The physiological requirements for this narrow repertory are few. To take a specific case, the male silkworm moth makes a sexual response only to a particular chemical substance, bombykol, emitted from abdominal glands of the female moth. The decision is made in an extremely simple manner by the smell receptors themselves. The male catches the bombykol molecules on 20,000 specialized sensory hairs deployed over his two feathery antennae. Each hair contains two cells that respond only to bombykol. They send their electric messages inward to the main antennal nerve and ultimately through connecting nerve cells to the brain. When enough of these signals are received by the brain, the male moth vibrates his wings for a short while, then flies out in search of the female.

When we turn to the olfactory behavior of mammals, a whole new level of brain organization is encountered. Male domestic cats urinate at chosen signposts throughout their territory in order to warn off potential rivals. The urine contains a few substances, including one chemically close to catnip, that cause a strong reaction in other males. However, tomcats do not simply respond to the odors of other males in the manner of most insects. They learn to recognize individual scent mixtures at the signposts of their rivals and to judge the whereabouts of the other cats by the degree to which the scent has faded. They then employ this information in deciding whether to remain, retreat, or change direction.

The greater subtlety of odor discrimination by the domestic cat and virtually every other category of its behavior has been obtained at the price of an enormous increase in the number of nerve cells. Instead of the hundred thousand to a million found in the silkworm moth, there are a billion or more in the cat. The human brain is even larger and more complex, containing approximately 100 billion nerve cells and an average of about 10,-000 connections per cell. About 20 percent of the brain is devoted to speech and language. This portion comprises widely separated portions of the cortex that cooperate swiftly and precisely during most events of communication and conscious thought. In the course of fetal development the brain adds nerve cells at the rate of hundreds of thousands every minute. The possibilities for error are enormous during this phase of rapid growth and in later life, when the full-sized brain must be maintained through the processing of millions of bits of information. It is not surprising that neurologists and psychiatrists have compiled a long catalog of genetic and environmentally induced defects and diseases, which together impair almost every aspect of mental activity—crippling, blinding, or disorienting their unfortunate bearers. As Seymour Kety has said, "The wonder is that for most people the brain functions effectively and unceasingly for more than 60 years."

All these intrinsic difficulties were somehow overcome during human evolution. Some extraordinary set of circumstances—the prime movers of the origin of mind—must have existed to bring the early hominids across the Rubicon and into the irreversible march of cultural evolution. The prime movers have been the subject of a great deal of speculation. One of the most intriguing explanations, suggested by Charles Darwin himself, is warfare. There is plenty of evidence of violent aggression among recent bands of hunter-gatherers, whose social organization most closely resembles that of primitive man. As Darwin said in *The Descent of Man*, the practice of each new art of war "must likewise to some degree strengthen the intellect. If the invention were an important one, the tribe would increase in number, spread, and supplant other tribes. In a tribe thus rendered more numerous there would always be a greater chance of the birth of other superior and inventive members. If such men left children to inherit their mental superiority, the chance of the birth of still more ingenious members would be somewhat better, and in a very small tribe decidedly better." To put Darwin's hypothesis in somewhat more modern form, one band would in time develop the capacity to ponder the significance of adjacent social groups and to deal with them in an intelligent, organized fashion. A band might then dispose of a neighboring band, appropriate its territory, and increase its own genetic representation in the population at large, retaining the tribal memory of this successful episode, repeating it, increasing the geographic range of its occurrence, and spreading its influence still further in the population.

Mathematical analyses of this pattern of tribal dominance and genocide indicate that warfare has the capacity to accelerate evolution. Aggressiveness may well be the dark underside of the human intellect. But there is no compelling reason to identify it as the prime mover that created the capacity for culture. Warfare is common among groups of hyenas and chimpanzees, and it almost certainly occurred in other highly social, large-brained animals during past eras.

Perhaps the prime mover can be found instead in sexual competition. As a rule men make elaborate efforts to obtain more than one sexual partner. The great majority of economically more primitive societies are polygamous, and the number of wives is generally regarded as a measure of masculine success. The anthropologist Robin Fox has argued persuasively that the competition for women was a powerful stimulus to the evolution of intellect. Sexual selection, like that in many species of monkeys and apes, probably took the form of both artful displays toward females and aggression toward rival males. But the aggressive element could not have been too overt because primitive men also depended to an unusual degree on cooperation. The competition is likely to have consisted instead in comparisons of hunting prowess, leadership, skill at toolmaking, and the other obvious attributes that contribute to success of the family and the band. Another anthropologist, Sarah Blaffer Hrdy, has added the female side of this equation: skills in discrimination, judgment, and the most subtle forms of communication—and a remarkable repertory of aggressive techniques. In the sexual arrangements of the nearly egalitarian hunter-gatherer societies, women have probably always been equal intellectual competitors.

The overall idea is very appealing: the engine of sexual competition drives the evolution of the brain independently of the environment. This notion can also be traced to Darwin, who suggested that sexual selection is a very special and autonomous form of natural selection. No doubt sexual competition did play a role, and perhaps a fundamental one, in the evolution of the human mind. But it is not likely to have been the prime mover: sexual competition is too widespread a phenomenon in the animal kingdom. It occurs in diverse forms among both solitary and social species, producing variously gorgeous, powerful, and cunning males and, under appropriate conditions, equally beautiful and formidable females.

So the search for the elusive prime mover must turn in new directions. A physiologist, K. R. Fialkowski, has nominated heat stress as a key factor. When early men began hunting on the African savannas (according to this hypothesis), they lacked adequate cooling mechanisms for the brain. An excessively high blood temperature damaged nerve cells and impaired mental ability. The species responded by evolving larger brains with less densely packed and hence more redundantly acting nerve cells. Another biologist, Valerius Geist, has drawn the opposite conclusion. He shifted the locale of human cranial evolution out of the tropical savannas and into the cool, rapidly shifting temperate zone during the ice ages. The most challenging environment was the "rich, diverse, and demanding periglacial environments at the fringe of glaciers in Eurasia." Here new habitats continuously opened up, and human populations grew by repeated demographic bursts in the midst of challenging conditions. Evolution was therefore maximally accelerated—not in Africa where mankind began, but to the north, where it subsequently spread.

In trying to judge these diverse ideas in a neutral manner, one cannot avoid the disconcerting feeling that there may be truth in all of them. The explanations are for the most part logical and consistent with the evidence. Any one or a combination of the selection pressures they assume could be a principal force in human mental evolution, and all deserve further study. Yet none identifies the conditions unique to the origin of man.

So let us go back to the very beginning of the true human line and have another look at *Homo habilis*, our diminutive and presapient ancestor. Perhaps we already have the clues to the breakthrough in the fragments of evidence lying before us. *Homo habilis* had a peculiar set of traits that adapted it to a certain niche in the African savanna. These adaptations, which evidently advanced it to a greater mental capacity than had existed in any previous species, are individually unimpressive but unique in combination: a moderately large and complex brain by nonhuman standards, the chance heritage of an Old World primate ancestry; an Old World primate social organization, another legacy that included the close association of kin and a mixture of aggressive and cooperative behavior; bipedal locomotion; hands freed from locomotion and thus made available for other uses; and not least, the killing of animals for food. All of these specializations were put to the service of a hunting and gathering life on the savanna. The early *Homo* were small and not very fleet, at least not in comparison with big cats and antelopes. They lacked fangs, sharp hooves, horns, or claws. It seems to follow closely that a premium was placed on intelligent, cooperative action and on the use of tools. Food could be carried long distances and shared with others, and labor could be divided between those who foraged and others who protected the young. And this opportunity placed a still higher premium on sophisticated communication and social rules. In short, the combination of an already sizable brain, a primate social organization, and free hands overcame the resistance to advanced cognitive evolution that had shackled the living world during the previous two billion years.

If this interpretation is correct, the long-sought prime movers nudged the human line forward until gene-culture coevolution ignited and became a self-sustaining reaction. The driving force then propelling the human species from the *Homo habilis* to the *Homo sapiens* level was not war, sex, climate, or hunting on the savanna, but gene-culture coevolution. Its products and its instruments are one and the same: advanced tool use, language, and long-term memory. Cultural innovations acted as a new class of mutations that accelerated evolution and pushed the species forward to its present genetic position.

This view of what happened can be made clearer with the use of a geological metaphor. The origin of mind was like the radical transformation of the highest peak of a tropical mountain range. In its history and composition the eminence does not differ in any fundamental way from the foothills and peaks around it. But because it was located in just the position where the forces of crustal uplift pushed it a little bit higher, it acquired snow and ice and unique forms of alpine life. A threshold was waiting; a small quantitative change then resulted in the abrupt creation of a new world.

Because high intelligence and culture appeared only once in the entire history of life, it is not possible to say whether the peculiar and adventitious prime movers that altered primitive men are the only ones that could place any species at any time within the reach of gene-culture coevolution. What we do suggest is that in the human case coevolution took the old primate traits that happened to be present and transmuted them into the epigenetic rules of mental development unique to man. Some traits were no doubt erased, with the bias being neutralized or reversed. Others, such as reciprocity and cooperativeness, were strengthened. New epigenetic rules were added in the more sophisticated activities of memory, reasoning, and valuation. But the direction of the coevolution was unavoidably set by the biology of the species that gave rise to modern man—that is, it was prefigured in *Homo habilis* and *Homo erectus*. The mind is a chimaera of free will and idiosyncratic survival devices reminiscent of the ancestral hominids. Out of all of the thousands of possibilities that existed and might have come to fruition, the regnant species on earth therefore happens to have a human nature.

Prometheus may now receive his proper due as mythic hero. In every respect, through unprecedented triumph and agony, he served the needs of human ambition. He stole fire from the gods and gave its creative power to the human race. He suffered dire punishment for that contemptuous act but remained defiant and confident of the rightness of his deed. In literal history Prometheus probably began in earliest Greek mythology as a fire god. He was later eclipsed by the oriental Hephaestus but remained popular as the supreme trickster and the patron saint of the proletariat. Because fire is the material basis of culture, linked as in the *Rig Veda* with widsom to become "the friend of man, the immortal among mortals," the Promethean gift was the supreme event in the Hellenic narrative of the origin of man. The

Prometheus of Aeschylus could rightfully boast: "Who was it but I who in truth dispensed honors to these new gods? . . . One word will tell the whole story: all arts that mortals have come from Prometheus."

In our own lifetime this gift is being refashioned into the instrument of independence. Man has magnified self-reflection into the centerpiece of existence; we can examine history and plan into the distant future, even our genetic future. Through culture we have literally altered the form of organic evolution and can now take control, if we wish. But ours is also the most dangerous of all times. The key events in the transition have slipped out of phase. We are still too ignorant to be free, too dangerous to remain in genetic thralldom, and just possibly too vain and fearful to find a way out.

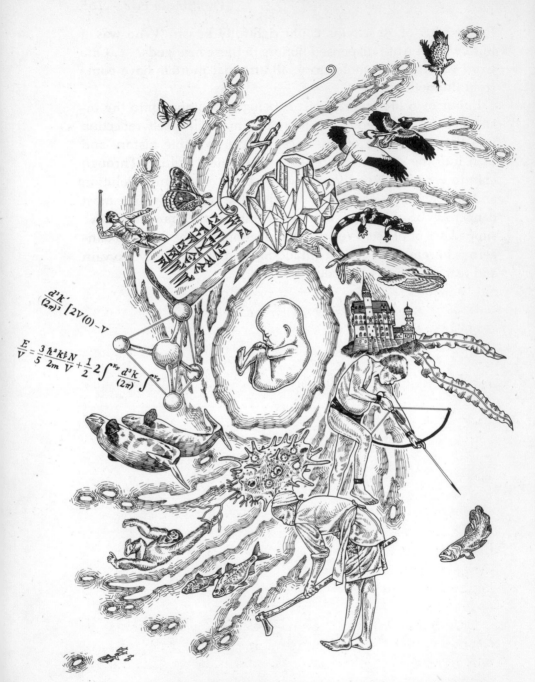

$$\frac{d^3k'}{(2\pi)^3}\left[2V(0)-V\right]$$

$$\frac{E}{V}=\frac{3}{5}\frac{\hbar^2k_F^2}{2m}\frac{N}{V}+\frac{1}{2}2\int^{k_F}\frac{d^3k}{(2\pi)}\int^{k_F}$$

The evolution of humankind is one thread in the seamless pattern of nature.

# Toward a New
# Human Science

THE FOSSIL DISCOVERIES in Africa have disclosed the evolution of the human line going back more than two million years. Anthropologists are familiar with the ancestral human beings scattered along the way: the contours of their face, the size and surface features of their brain, the way they walked, the very rocks they pressed in their hands, the animals they killed and the predators that hunted them in turn, the trees around them, their springs and summers.

But the bones and rocky matrix laid out on laboratory tables reveal only external features of early human life. We have argued that a different missing link now ranks as the central problem of evolutionary reconstruction: the human mind. It is by far the most complicated process that ever existed on Earth; and its generative organ, the brain, an inner universe of one hundred billion or more nerve cells, is the most complicated machine. Although in many ways the mind is the ultimate object of scientific inquiry, its evolutionary history remains a blank chapter in the textbooks.

Only during the quiet revolutions in the brain sciences and cognitive psychology of the past twenty years have we come fully to realize how little is known about the brain and mind. It was necessary first to break the iron grip of behaviorism, which restricted much of psychology to the study of stimulus and response and drew too simple a sketch of human behavior. At first the behaviorist philosophy operated effectively as a purifying force: it imposed tough new standards of experimental proof and erased the last traces of mysticism from psychology. But in the process it demoted mind to an illusory concept and depicted the brain as a virtual blank slate upon which experience is inscribed according to elementary laws. As these more radical strictures of behaviorism waned in influence, mental process—cognition—came to be seen as not only real but composed of a series of complicated steps by which information is acquired and behavior is decided. Psychologists discovered that the mind automatically classifies light into colors according to wavelength; spaces phonemes at certain intervals during the creation of language; prefers sugar, eyes, a smiling face, and bull's-eye designs; falls prey to phobias about snakes but seldom about knives; underestimates major risks; uses certain procedures of "fuzzy logic" based on prototypes when classifying objects; and so on through every other major category of perception and thought thus far put to serious study. Most and perhaps all these particularities have a biological basis, encoded into the location and responses of various sets among the billions of nerve cells. An exciting new world of phenomena has been opened up. It is now clear that the laws of psychology, including those governing the process of learning, are not elementary at all.

The importance of understanding this cognitive mélange cannot be overstated. Almost everything that human beings feel to be the truth and every act that we can perform in good conscience ultimately depend on it, because our interpretation of truth and value is fixed by how we believe the mind works. Cognition has become a burning issue even within the ranks of the theoretical physicists, a few of whom are convinced that the mind of the observer is somehow woven into the laws of subatomic particles. Beyond physics, philosophy itself can be said to consist largely of failed theories of the brain.

To understand cognition and the evolution of the mind is to

penetrate the mysteries of human nature and to put to the test powerful political and religious beliefs. Traditional Marxism, to take a prominent example, views the mind as biologically unstructured. In effect it denies the existence of a constant human nature. It conceives of history as the result of external forces, particularly economic change and class struggle, that precipitate revolution and move people toward a classless society. Marx himself, after vacillating a good deal on the subject, concluded that "it is not the consciousness of men that determines their existence, but their social existence that determines their consciousness." It follows that human nature, if it exists, is something transformed continuously by history. Thus a socialist utopia, or for that matter any kind of utopia, can be attempted without paying much attention to biology. Perhaps for this reason, a few (but by no means all) Soviet psychologists have chosen to stay huddled in the early twilight of Pavlovian learning theory, with little interest in psychoanalysis and genetics. The traditional Marxist view is contradicted fundamentally by the discoveries of rich structure in the operation of the mind and in the development of social behavior, much of which has nothing to do with socioeconomic forces. Whether researchers oriented toward Marxism can come to terms with this knowledge without abandoning their central metaphysical beliefs will be an instructive experiment in the sociology of science. Meanwhile, it is important to recognize that ideologies of both the left and right are threatened not so much by human nature as by the failure to confront it in an open and unbiased manner.

Scientists consider the mind to be one of the last frontiers of basic research. Perhaps it will prove the very last and greatest. It is already the center of an implosive convergence of disciplines that were previously unconnected: the brain sciences, computer science, psychology, linguistics, anthropology, ethology, genetics, neurophysiology, sociobiology, and, not least, philosophy. The hoped-for result will be a new human science, which accepts mankind as a product of evolution and might, given enough time, succeed in reinterpreting history as the interaction of biology and culture. And what will be involved is history in the fullest sense. It is often said that nothing makes sense except in the light of history, meaning cultural change over a few centuries. More accurately, nothing makes sense except in the light of organic evolution, which encompasses a tightly linked form of cultural and genetic change and spans hundreds of thousands of years.

At this point sociobiology has entered the scene as an entirely predictable intellectual development. Defined as the study of the biological basis of social behavior, it carries evolutionary theory into the previously un-Darwinized fields of psychology and the social sciences. Conventional sociobiology first addressed the genetic origins of such general behaviors as altruism, cooperation, sexual bonding, parental care, and aggression. It identified the special environmental conditions in which these and other phenomena evolve, the advantages they confer on the members of societies, and consequently their significance in an ultimate, biological sense. But the human condition is dominated by two qualities that cannot be handled by ordinary evolutionary theory and sociobiology: the human mind, operating with free will, and culture, which has created an astonishing diversity of behavior among different societies.

The proper account of mind and culture must depend on a theory of gene-culture coevolution, which is the understanding in modern scientific terms of the great circuit of causation that runs from the genes to brain architecture and the epigenetic rules of mental development, then to the formation of culture, and finally back to the evolution of the genes through the operation of natural selection and other agents of evolution. We have offered evidence suggesting that hereditary and environmental influences cannot be cleanly separated. They work together to create the whole brain, mind, and culture. At the next higher level, genetic and cultural evolution are also permanently linked, a circumstance that led to our choice of the expression *coevolution*. We suggest that the human species is the product of a unique episode of gene-culture coevolution that has accelerated the growth of the brain and mental capacity during the past two million years. So potent is the process that some of the genetic advance in symbolic reasoning and language could have occurred during the past 50,000 years or less. It might be continuing right into historical times.

The conglomerate of enterprises that make up human science are in the earliest stage of development. The brain remains a largely unexplored world; the circuitry of its hundred billion nerve cells has been no better mapped than the valleys of Mars. No one knows where or how memory is stored, or the physiological nature of consciousness. At the level of psychology the known epigenetic rules seem to be the mere fringe of a still undeciphered code. The translation of these rules into cultural patterns has been attempted with the aid of mathematical models only in several of the simplest, most tractable cases, such as brother-sister incest.

A guiding principle has nevertheless reemerged from the combined efforts that once inspired Comte, Spencer, and other nineteenth-century visionaries before dying from premature birth and Social Darwinism: that all of the natural sciences and social sciences form a seamless whole, so that chemistry can be unified with physics, biology with chemistry, psychology with biology, and sociology with psychology—all the way across the domain of inquiry by means of an unbroken web of theory and verification. In the early years the dream was bright. The link between physics and chemistry came relatively early, starting with rapid advances in the 1800s; the creation of physical chemistry was a triumph of scientific synthesis. The connection between chemistry and biology was contested well into this century and could not be completely confirmed until molecular biology matured in the 1960s. The bridge between biology and psychology is still something of an article of faith, in the process of being redeemed by neurobiology and the brain sciences. Connections beyond, to the social sciences, are being resisted as resolutely as ever. The newest villain of the piece, the embattled spearhead of the natural-science advance, is sociobiology.

Sociobiology has been accused of reductionism, the practice of explaining an organism, a society, or some other complex entity entirely by the parts that compose it. Sociobiology *is* reductionistic in spirit, but it can also be said that the procedure of breaking down complicated systems into manageable components, in other words analysis, is only half the standard method of science. The other half is synthesis, in which the relation of the parts is also laid bare and the whole system then reassembled either by direct experimentation or by theoretical simulation with mathematical models. To identify an unknown carbohydrate, for example, the molecule is often first broken down into smaller, identifiable fragments. Then, if a final test is needed, the substance is synthesized from smaller molecules taken from the laboratory shelf. In the study of a brain or society, the events deemed to be of greatest importance are examined in detail and their parts teased apart and identified. Then they are recreated— not by a literal physical union as in the case of the carbohydrate molecule, but by logic and mathematical operations. Synthesis is both the appropriate second enterprise of science and a powerful means of verifying analysis. The ultimate procedure of science is thus an act of creation, a reconstitution of the real world in vivid and efficient form.

It is often claimed that the analytic-synthetic procedure of the natural sciences will be defeated by history because historical particularities are so complicated and unpredictable that they overwhelm everything else in the explanation of human behavior. But these are merely claims. At present there is no scientific proof that the essential qualities of a Charlemagne or the Quattrocento cannot emerge from a sophisticated theory of gene-culture coevolution applied to history. And though such an advanced formulation is a task for the future, a great deal can even now be deduced about societies and history. Theoretical studies can incorporate many of the distinctive features of any system, whether inanimate or living, by specifying the conditions under which the system began and under which it developed. The simplest of worlds already treated by science, such as the trajectory of a missile, the break of a wave, or the cyclic course of an epidemic, can be predicted by a knowledge of the appropriate governing laws plus a specification of the starting conditions and key features of the environment in which the particular event is about to occur. Although the human mind and culture are vastly more complicated than these elementary phenomena, there is no reason to expect that they cannot be treated and more deeply understood by a comparable stepwise procedure.

The critics argue back: history is neither a breaking wave nor an influenza epidemic. The importation of biology into the social sciences must fail because it cannot account for the purpose and ultimate goals in the human mind. Meaning is beyond the reach of evolutionary models. This objection is persuasive in sound, but it is no longer true. Computer science has demonstrated that sophisticated goals can be built into machines. The functionalistic approach to the mind-body problem has further shown that goal-oriented behavior in machines and brains can be closely compared. Finally, the theory of evolution by natural selection has revealed the means by which purpose can evolve in brains as a mechanism furthering survival and reproduction. It follows that the human mind can conceive of a biological device that creates purpose—and from purpose, meaning.

The chorus of critics responds: no, something is drastically wrong. This mechanistic conception of the human mind will be disastrous even if it is just accepted on faith—it takes away free will. If the genes program the brain and orchestrate the assembly of the mind through the epigenetic rules, then the genes are the ultimate masters, and "we"—the minds now engaged in this act of communication—are not really responsible for our thoughts and actions. It would seem to follow that the more we excavate the roots of human nature, the more thoroughly we will come to understand the genetic commands. The result will be a progressive shrinking of freedom and personal responsibility, a confusion of good and evil. Perhaps certain kinds of knowledge can be dangerous after all. At the very least, ethics must never be subordinated to biology.

In one sense, the dangerous-knowledge argument is the most damaging that can be brought against sociobiology or any other materialist interpretation of the origin of the mind, since it is directed at what everybody, expert and layman alike, holds to be the most precious human legacy: freedom, choice, selfhood, spirit, and hope. But this line of reasoning is fallacious. A further examination of the problem reveals that, quite to the contrary, a deep knowledge of human nature can only increase free will, not diminish it.

Here is the essence of that argument. All of our behavior is indeed predestined to the degree that we have deeply ingrained goals and principles that organize our daily lives. The free choices made are for the most part thoughts and actions put to the service of these internal guides. Some scholars believe that goals and principles are acquired almost wholly from the surrounding culture, with no genetic biasing. But even if this were true, the individuals would still be determined by forces external to themselves. In this case they would be programed by their culture. Cultural determinism can be as much of a straitjacket as genetic determinism.

True free will and spiritual independence are nonexistent in people guided solely by the norms of their culture. Even the bright individuals who grasp this fact and challenge the norms are still not free. They are merely reaching for new sets of goals and principles, an endeavor in which they will be skillfully led by the deep impulses and feelings prescribed by their genes. Most rebels are people who merely try to exchange one set of chains for another, without really knowing why. In contrast, a scientific understanding of human nature and the process of gene-culture coevolution can provide some measure of intellectual independence from the forces that created us. It can enhance true free will. Real freedom consists of choosing our masters by a procedure that allows us to master them.

The scientific study of human nature also seems to be the appropriate way, if any exists, to create value-free social sciences. When the roots of ethics and motivation are fully exposed, polit-

ical science, economics, and sociobiology can be more easily un-
coupled from the genetic and cultural biases of the specialists
who originate them. Ethics is the keystone, the override switch
that forestalls abuse and tyranny. But ethical philosophy should
no longer operate outside the boundaries of science. Leaving it to
the vagaries of genes and culture and the unaided intuition of
great thinkers can be dangerous, as history has amply shown.
Because moral judgment is a physiological product of the brain,
it too can be greatly assisted by the new, human-focused science.

For a practical example of this approach, let us return a final
time to the dilemma of incest. No act has intruded more tragi-
cally into the canons of art and religion, none seems more threatening to the social fabric. Oedipus tore out his own eyes to atone
for it. And yet the suggestion is sometimes made by modernists
that, since the incest taboo was created by culture, it can be un-
done by culture. Perhaps, this cheerful reasoning goes, the taboo
is merely the last surviving barrier to the sexual revolution, and
upon dismissal it will be found to have been no more than a
communal delusion, not unlike the failed curse of a sorcerer.

If we could float down like disembodied spirits and ponder the
matter free of prejudice and scientific information, this interpretation might at first seem to have great value. There is a certain
abstract beauty in the thought of an erotic union between
brother and sister. Possessed of an identical ancestry, half their
genes the same by reason of common descent, and with the
memories of a thousand sensitive moments shared during the
formative years of their childhood, they are as close as two
young people can be. To unite sexually might be near perfection,
an act of special beauty. It combines self-love and the joy of fus-
ing with another person. It encloses and protects the genetic heritage and material wealth of the family. It offers just about everything that human beings wish for themselves—sexual pleasure,
sharing, deep intimacy, security, and not least, simplicity.

So why is it a sin? The cultural determinists would say, well,
because that is what people have learned from their culture. It
seems to follow that we can go ahead and dispose of the taboo, in

other words experiment with changing the culture, and see if that improves the quality of life. A few venturesome families might even actively encourage brother-sister marriages. But the vast majority of the population would certainly feel otherwise, and strongly. In their guts they know incest is wrong. For them removing the barrier would be obscene in the oldest sense of the word (*obscenus*, filthy), the worst form of sacrilege. The modernists could reply: that is what you used to say about premarital sex and birth control. At least give us the freedom to practice incest on our own. Remove the repressive laws and, above all, spare us your hypocritical posturing.

Disembodied spirits might continue this debate forever. But it can be settled quickly with the aid of sociobiology. It is natural—in the full, biological sense—for people to be opposed to brother-sister incest. As we explained earlier, an epigenetic rule exists that causes the great majority of individuals to reject this form of incest in favor of sexual attachments outside the family. The rule is so strong and pervasive across history and societies that it can be reasonably supposed to have a genetic basis. Of course individuals continue to exercise free will in the matter. They can think about the problem at any depth down to the epigenetic rule and the genes and with due attention to their own special circumstances. Then they can decide. Most likely they will resist incestuous acts because the epigenetic rules have nudged their minds into forms that are much more likely to receive deep emotional reward from the practice of sex outside the family in which they were raised.

But the genes can be tricked. The epigenetic rule is the following: children reared closely together during the first six years or so are not sexually attracted to one another at maturity. Knowing this, any society wishing to promote incest can simply arrange to have brothers and sisters raised apart, then encourage them—by law, art, music, religious dictum—to marry. Individuals would now in all probability prefer to marry siblings, and in so doing they would continue to exercise free will. But a superordinate form of free will has also been exercised by other members of the society to circumvent the genes and raise the percentage of incestuous marriages. The trick has been achieved by understanding the biological basis of incest avoidance and exerting considerable effort to subvert it.

This feat of social engineering, if it were ever to be undertaken, is comparable to the procedures of biochemical engineering used to correct genetic defects. The condition called phenylketonuria (PKU) is typical of many genetic diseases in that it is caused by a single recessive gene (hence it is expressed only when in double dose) and is relatively rare, occurring in one infant out of every ten thousand. When an infant has a double dose of the PKU genes, it is unable to utilize the amino acid phenylalanine, a common constituent of protein diets. Abnormal amounts of phenylalanine accumulate in the blood, poisoning the system and causing irreversible brain damage if not corrected within several months after birth. PKU can be prevented by reducing the amount of phenylalanine in the diet.

Neither PKU nor incest avoidance is caused entirely by genes. Both are created by an interaction of heredity and environment. Nevertheless, in almost all environments in which human beings live, PKU genes will lead straight to PKU, and the genes presumed to underwrite sexual preference will lead straight to incest avoidance. Only by understanding the nature of the heredity-environment interaction is it possible to select the precise environment that reverses the usual response—that is, less phenylalanine as a means of preventing PKU, and raising brothers and sisters apart as a means of increasing the frequency of incest.

So individuals make choices, societies as a whole make another set of choices, and the societal choices in turn influence the individual choices. The individual expresses himself, the society practices social engineering, and both enjoy free will. Where does ethics fit in? Individuals are behaving in an ethical manner because they make what they believe deep inside to be the correct choice. But the society can, if it possesses appropriate knowledge, try to reverse those deep feelings, even when they are powerfully steered by the genes. In other words, ethical premises are not immutable or transcendent. We like to think they were handed to us on stone tablets, but they can be changed at will.

But whose will? And in obedience to which superordinate rules? If society were to choose not to circumvent the PKU gene, it would end up with a larger number of handicapped and dead children. If it were to choose to circumvent the genes under-writing the incest taboo, it would increase the rate of inbreeding, multiply genetic defects (including PKU), and end up with a much larger number of handicapped and dead children.

Now we have arrived at consequences on which almost every-one will agree. Almost no one believes that individuals should act in a way that deforms and kills children. Yet this too is an-other gut feeling like the one that condemns incest in the first place. If the genes can be circumvented in a way that makes peo-ple prefer incest, and the ethical premise changed at that level, why should it not be possible to trick the genes again and pro-duce minds that tolerate or prefer a greater number of deformed children? The idea seems even more preposterous (to us, right this minute) than the promotion of incest, but the result is theoretically possible. People still make choices in matters of life and death. They are not hardwired automata of virtue. Child abuse and murder are relatively common, and once in a while societies favor infanticide or wage genocidal wars. With appro-priate knowledge and techniques of genetics and cognitive psy-chology, it could be possible to create a society that not only ap-proves of physical handicaps and early death, but feels deeply that this must be right—that God wishes it so, or the immutable laws of history dictate it—and invents songs, religions, and so-phisticated literature to justify the new ethical code. (Intellectuals might even write Swiftian essays in *The American Spectator* justi-fying the practice on the grounds that it cleanses the species of harmful genes. Beautifully composed rebuttals would quickly follow in *The New York Review of Books* denying that genes exist.) Some kinds of animals are in fact programed to eat a few of their young or to commit suicide during the process of reproduction.

If such creatures could think and speak, they would heartily approve of this murderous ethos.

So we suggest that moral reasoning is based on the epigenetic rules that channel the development of the mind. Such reasoning appears to be ultimately dependent on the genes as well as on culture and self-conscious decision. But the rules only bias development; they do not determine ethical precepts or the necessary decisions in a fixed manner. They still require that a choice be made, and in this sense they preserve free will. However, a sufficient knowledge of the genes and mental development can lead to the development of a form of social engineering that changes not only the likelihood of the outcome but the deepest feelings about right and wrong, in other words the ethical precepts themselves.

If this biological interpretation is correct, moral reasoning can be assisted by consulting the ethical precepts through a method of regress. When you make a decision—say incest versus outbreeding—you examine your feelings about what is right or wrong. To put the matter in somewhat more mechanical terms than is customary in ethical philosophy, you play out scenarios in the conscious mind and note the emotions connected to the alternative course of action. You think and feel—and then choose the most satisfying scenario, knowing that you have done the "right" thing. But society at large—the nation, the corporation, the family, or any other authority-bearing unit—has the power to alter the rules in this first line of moral reasoning. The decision it faces is not the behavioral act but the choice of ethical precepts that affect the class of such acts. To make that choice it falls back on the second line of moral reasoning. The members of the group understand that aversion to incest is a visceral feeling that can be reversed with appropriate techniques.

But before a decision is made, the social engineers must examine the consequences of reversing the ethical precept. They find that incest produces defective children. A new ethical precept is now confronted, one that chooses between normal and defective children. This precept might also be reversed in favor of defective children if sufficiently heroic measures were employed. It is biologically based and no more intrinsically sacred than the incest taboo; the emotional feelings against it are not much stronger. But at last we have come close to rock bottom. Everyone, starting with the experts, is likely to reject such an ethical mutation. The cost of the change is first of all great in both time and energy. More important, pain and suffering will ensue, and the very survival of the society might be put at risk. These considerations touch a layer of ethical precepts that can be altered themselves but that few will want to change. As a consequence, the existing aversion to incest is likely to be preserved; no community-wide support will be given to childrearing practices or to any process of biology and mental development that circumvents it.

This approach to ethics and political decision can be called a "top-down" analysis. We explore the consequences stepwise, starting with the immediate set of choices and the techniques required to make each choice acceptable or even preferred. This information discloses deeper, more hidden sets of choices. It calls for additional methods to render each of the second line of choices ethically acceptable. Finally, when the analysis has cut deeply enough, a consensus can be reached that is more extensive and secure than a consensus reached by means of dogma or unaided intuition. A decision at the uppermost level, say incest or outbreeding, can be made with a fuller knowledge of all the consequnces that will follow at lower levels.

The top-down analysis suffers from the fact that it treats each ethical problem in isolation. The ethical philosopher typically starts with real-life situations and tries to define the deeper mental and biological processes that are relevant. A complementary and in some ways more satisfying approach is attainable, which might be called the "bottom-up" analysis. This method starts with the deeper processes and works back up to the various real-life choices they may effect. Some epigenetic rules are powerful and highly constraining. They produce a narrow range of attitudes and behavior across all societies and can be altered only with difficulty. Others can be played upon to produce a vast array of effects on culture with relatively little psychological strain on individuals. Some of the epigenetic rules regulate the development of ethical knowledge and intuition. When their inner mechanism is finally blueprinted, society will be in a position to identify those culturally transmitted ethical choices most likely to penetrate the developmental barriers and hence to endure. The search for these belief systems constitutes the scientific equivalent of ethical knowledge "out there" beyond the vagaries of culture. In this endeavor, we may find that certain sexual mores and forms of mother-infant bonding, though rigid, are also congenial in all their ultimate consequences. On the other hand, qualities of leadership and aggressive behavior are likely to prove relatively plastic, with only a tiny percentage of the outcomes consistently favorable. We may discover hitherto unsuspected constraints and capacities among the many processes of cognitive development remaining to be explored. The chief advantage of bottom-up analysis is that it comes closer to the set of fundamental ethical imperatives. It can also provide a fuller and more useful picture of what human beings are able to do.

We have carried this exercise in moral reasoning to an extreme to illustrate several principles that we believe are inherent in human science. They can be summarized as follows:

- All domains of human life, including ethics, have a physical basis in the brain and are part of human biology; none is exempt from analysis in the mode of the natural sciences.
- Mental development is more finely structured than has been generally appreciated in the past; most or all forms of perception and thinking are biased by processes in the brain that are genetically programed.
- The structure in mental development appears to have originated over many generations through a specialized form of evolution (gene-culture coevolution), in which genes and culture change together.

- The biases in mental development are only biases; the influence of the genes, even when very strong, does not destroy free will. In fact the opposite is the case: by acting on culture through the epigenetic rules, the genes create and sustain the capacity for conscious choice and decision.
- The predispositions originate from an interaction of particular sets of genes and the environment; they can be altered in a precise manner if the appropriate information about them is available.
- Ethical precepts are based on the predispositions, and they too can be altered in a precise manner.
- One result of a strong human science might be the creation of a sophisticated form of social engineering, one that touches the deepest levels of human motivation and moral reasoning.

These propositions bring into sharper focus the main question of ethical philosophy: do moral codes exist in the absence of organic evolution? If there are such guideposts, the is/ought distinction is preserved. What the human species is at each stage in its evolution cannot be confused with what it ought to be. But if there are no such guideposts, the distinction does not exist.

A few modern philosophers, in recognizing this dilemma, have argued that the evolutionary explanation of human ethical values cannot by itself disprove the existence of absolute values outside the mind. Human beings may well be in the process of tracking the external truths by both genetic and cultural means. Just as a primitive ability to count objects and form abstract concepts has led to the discovery of elaborate mathematical theorems far beyond the needs of human survival, further moral reasoning could uncover ethical precepts that hold true for any genetic constitution that the human species may choose for itself in future generations. Once discovered, such adamantine truths can then serve as the lodestar of further cultural and genetic change.

But the philosophers and theologians have not yet shown us how the final ethical truths will be recognized as things apart from the idiosyncratic development of the human mind. In the meantime, by appealing to the core principles of neurobiology, evolutionary theory, and cognitive science, practitioners of a new human science can reach a deeper understanding of why we feel certain courses of action to be intrinsically correct. They can help us to understand why we have moral feelings. For now, though, the scientists can offer no guidance on whether we are really correct in making certain decisions, because no way is known to define what is *correct* without total reference to the moral feelings under scrutiny. Perhaps this is the ultimate burden of the free will bequeathed to us by our genes: in the final analysis, even when we know what we are likely to do and why, each of us must still choose

The challenge to science and philosophy to solve this dilemma is very great—in our opinion, there is none greater. Society, through its laws and institutions, already regulates behavior. But it does so in virtual blind ignorance of the deep reaches of human nature. By relying on moral intuition, on those satisfying visceral feelings of right and wrong, people remain enslaved by their genes and culture. Their minds develop along the channels set by the hereditary epigenetic rules, and while they exercise free will in moment-by-moment choices, this faculty remains superficial and its value to the individual is largely illusory. Only by penetrating to the physical basis of moral thought and considering its evolutionary meaning will people have the power to control their own lives. They will then be in a better position to choose ethical precepts and the forms of social regulation needed to maintain the precepts.

Social engineering has the potential of profoundly altering every part of human behavior. It will not always affirm the status quo, as in the case of incest avoidance. Some very human propensities, which may have been of great adaptive value in the stone age, are now largely self-destructive. The most virulent of these, aggression and xenophobia, can be blunted. Other equally human propensities for altruism and cooperation might be enhanced. The value of institutions and forms of government can be more accurately judged, alternative procedures laid out, and steps cautiously suggested. Economists and corporate planners, once aware of the facts of human nature and measuring more than material transactions, should be able to devise more effective policies.

Close self-examination and the planned manipulation of values can be a distasteful exercise. But in a world growing steadily more complicated and dangerous, the alternatives are not promising. A society that chooses to ignore the existence of the innate epigenetic rules will nevertheless continue to navigate by them and at each moment of decision yield to their dictates by default. Economic policy, moral tenets, the practices of child-rearing, and almost every other social activity will be guided by inner feelings whose origins are beyond comprehension. Such a society cannot effectively challenge the ancient hereditary oracle dwelling within the epigenetic rules. It will continue to live by the "conscience" of its members and by "God's will." Such an archaic procedure just might, by fantastic good fortune, lead in the most direct and untroubled manner to a stable and wholly benevolent world. More likely, it will perpetuate conflict and continue to drag humanity relentlessly along what is at best a tortuous and agonizing path.

On the other hand, the deep scientific study of the epigenetic rules will call the oracle to account and translate its commands into a precise language that can be understood and debated. People who know human nature in this way are more likely to agree on universal goals within the constraints of that nature and recognize absolute ethical truths, if such can be shown to exist. And though societies cannot escape the inborn rules of epigenesis, and would lose the very essence of humanness if they even came close to succeeding, they can employ knowledge of the rules to guide individual behavior and cultural evolution to the ends on which their members may someday agree.

# Notes

## The Fourth Step of Evolution

viii   The body form of *Homo habilis* shown here and in later figures follows the reconstruction by Jay H. Matternes, as presented for example in Richard Leakey, "Skull 1470," *National Geographic*, 143: 819–829 (June 1973).

3   Two of the most original and readable recent descriptions of early human evolution are by Alan Walker and Richard E. F. Leakey, "The Hominids of East Turkana," *Scientific American*, 239(2): 54–66 (August 1978); and Donald C. Johanson and T. D. White, "A Systematic Assessment of Early African Hominids," *Science*, 203: 321–330 (1979). The phylogenetic account we have given here, which is one of several competing evolutionary trees, follows the authoritative analysis by John E. Cronin et al. in "Tempo and Mode in Hominid Evolution," *Nature*, 292(5819): 113–122 (1981).

3   A technical review of more recent fossil evidence of the early prehuman apes is given by David Pilbeam et al. in "New Hominoid Primates from the Siwaliks of Pakistan and Their Bearing on Hominid Evolution," *Nature*, 270(5639):689–695 (1977). The African habitats of the prehuman ancestors are described by Peter Andrews, "Hominoid Habitats of the Miocene," *Nature*, 289(5800): 749 (1981). The habitats of early man are considered by (among others) Charles R. Peters in "Toward an Ecological Model of African Plio-Pleistocene Hominid Adaptations," *American Anthropologist*, 81(2): 216–278 (1979).

4   The significance of the distinguishing anatomical features of modern human beings and ancestral man are very well analyzed by Bernard Campbell in *Human Evolution: An Introduction to Man's Adaptations,* 2nd ed. (Aldine Publishing Co., Chicago, 1974).

5   For an analysis of the brain structure of early man and the man-apes, see R. L. Holloway, "Early Hominid Endocasts: Volumes, Morphology, and Significance for Hominid Evolution," in R. H. Tuttle, ed., *Primate Functional Morphology and Evolution* (Mouton, The Hague, 1975), pp. 391–415; and Dean Falk, "Hominid Brain Evolution: The Approach from Paleoneurology," *Yearbook of Physical Anthropology,* 23: 93–107 (1980).

5   Probably the best attempt to reconstruct the early evolution of language is by Sue Taylor Parker and Kathleen Rita Gibson in "A Developmental Model for the Evolution of Language and Intelligence in Early Hominids," *Behavioral and Brain Sciences,* 2(3): 367–408 (1979).

7   We speak of the four great steps of evolution as being at billion-year intervals only in the loosest sense. The earliest known microorganisms date to about 3.4 billion years before the present; see E. S. Barghoorn, "Aspects of Precambrian Paleobiology: The Early Precambrian," in Karl Niklas, ed., *Paleobotany, Paleoecology and Evolution* (Praeger, New York, 1981), pp. 1–16. The date of origin of eukaryotic cells, the second step, is very uncertain. There appear to be no such structures in the diverse and relatively well-preserved microflora of the Gunflint Chert, which is approximately two billion years old (Barghoorn). Some cells may be present in other deposits dating to 1.4 billion years, although one billion years is a safer estimate as the approximate time of origin. The earliest multicellular organisms with relatively complex, eukaryotic tissue appeared about 700 million years ago. The interval between the first eukaryotic cells and these organisms might have been in fact quite short, but an estimate cannot be made until the middle to late Precambrian record is more fully explored. A. H. Knoll and E. S. Barghoorn, "Precambrian Eukaryotic Organisms: A Reassessment of the Evidence," *Science,* 190: 52–54 (1975).

8   An excellent recent account of the history of the search for human fossils is by John Reader, *Missing Links: The Hunt for Earliest Man* (Little, Brown, Boston, 1981).

11  The subject of division of labor in animal societies including monkeys and apes is reviewed by Edward O. Wilson, *Sociobiology:*

*The New Synthesis* (Belknap Press of Harvard University Press, Cambridge, 1975).

11    The most detailed argument for the relation between upright posture and enhanced sexuality in the division of labor is given by C. Owen Lovejoy, "The Origin of Man," *Science*, 211(4480): 341–350 (1981). A popular account of these ideas is given by D. C. Johanson and Maitland Edey in *Lucy* (Simon and Schuster, New York, 1981). Opposing arguments are summarized in "Lucy's Husband: What Did He Stand For?," by Sarah Blaffer Hrdy and William Bennett, *Harvard Magazine*, July–August 1981, pp. 77–9, 46.

11    The evolution of human sexuality has been given new and sophisticated treatment by Donald Symons, *The Evolution of Human Sexuality* (Oxford University Press, New York, 1979), and Sarah Blaffer Hrdy, *The Woman That Never Evolved* (Harvard University Press, Cambridge, 1981).

11    Murder among the !Kung was discussed by Richard B. Lee in a talk, "!Kung Bushman Violence," at the annual meeting of the American Anthropological Association, November, 1969.

11    The evolution of human food sharing is examined by Irenäus Eibl-Eibesfeldt, "Human Ethology: Concepts and Implications for the Sciences of Man," *Behavioral and Brain Sciences*, 2(1): 1–57 (1979). The role of diet and food sharing in ritual is delightfully described by Peter Farb and George Armelagos in *Consuming Passions: The Anthropology of Eating* (Houghton Mifflin, Boston, 1980). Food exchange as a greeting ritual in children is noted by Parker and Gibson, "A Developmental Model."

13    An excellent and authoritative examination of the concept of mind is given by Donald R. Griffin and twenty-four coauthors in *Animal Mind—Human Mind*, Life Sciences Research Report 21, Dahlem Konferenzen (Springer-Verlag, 1982). These scientists also consider the various methods for reconstructing cognitive evolution.

15    The earliest stone tools were discovered by John W. K. Harris of the University of Pittsburgh, as described in *Science News*, 119 (7 February): 83–84 (1981).

15    In *Genes, Mind, and Culture* (Harvard University Press, Cambridge, 1981) and elsewhere we have employed a broad definition of culture as the sum of all of the artifacts, behavior, institutions, and mental concepts transmitted by learning among members of a society, and the holistic patterns they form.

15    The use of red ochre by *Homo erectus* has been convincingly inferred from fragments and powder at a Czechoslovakian site by J. Fridrich, "Ein Beitrag zur Frage nach den Anfängen des künstlerischen und Äesthetischen Sinns der Urmenschens (vor-Neanderthaler, Neanderthaler)," *Památky Archeologické*, 67: 5–30 (1976).

15    The Neanderthal buried at Shanidar was reported by Ralph S. Solecki in "Shanidar IV, a Neanderthal Flower Burial in Northern Iraq," *Science*, 190: 880–881 (1975).

15    The evidence of earliest writing, consisting of regularly arranged sequences of scratches on rocks and bones, was given by Alexander Marshack, "Upper Paleolithic Symbol Systems of the Russian Plain: Cognitive and Comparative Analysis," *Current Anthropology*, 20(2): 271–311 (1979).

16    The evolution of early agricultural societies is reviewed by Gerhard and Jean Lenski, *Human Societies: An Introduction to Macrosociology*, 3rd ed. (McGraw-Hill, New York, 1978).

16    The earliest symbolic writing was documented by Denise Schmandt-Besserat, "Decipherment of the Earliest Tablets," *Science*, 211: 283–285 (1981).

16    An excellent short account of the explosive growth of science is given by Derek de Solla Price in *Science Since Babylon*, enlarged ed. (Yale University Press, New Haven, 1975).

16    The rate of theorem production in mathematics is noted by P. J. Davis and R. Hersh in *The Mathematical Experience* (Birkhäuser, Boston, 1980).

17    The projection of computer capacity is due to Robert Jastrow, "The Post-Human World," *Science Digest*, 89(January–February): 89–91, 144 (1981).

17    Pierre Teilhard de Chardin, *The Phenomenon of Man*, trans. Bernard Wall, intro. Julian Huxley (W. Collins Sons, London, 1959).

18    The estimates of species extinction are from Norman Myers, *The Sinking Ark* (Pergamon Press, New York, 1979). See also Paul and Anne Ehrlich, *Extinction* (Random House, New York, 1981).

19    Recent works stressing the autonomy of the humanities include Kenneth Bock, *Human Nature and History: A Response to Sociobiology* (Columbia University Press, New York, 1980); W. I. Thompson, *The Time Falling Bodies Take to Light: Mythology, Sexuality, and the Origins of Culture* (St. Martin's Press, New York, 1981); and John

Bowker, "The Aeolian Harp: Sociobiology and Human Judgment," *Zygon*, 15: 307–333 (1980).

19    Our conception of gene-culture coevolution was first presented in "Translation of Epigenetic Rules of Individual Behavior into Ethnographic Patterns," *Proceedings of the National Academy of Sciences*, 77(7): 4382–4386 (1980); and elaborated in *Genes, Mind, and Culture*.

## The Sociobiology Controversy

23    There are now five technical journals devoted primarily or exclusively to sociobiology, of which the first three listed below have come into existence since 1975:

*Behavioral Ecology and Sociobiology* (Springer Verlag, New York). Contains articles on animal behavior and social organization, with emphasis on experimental and field studies as opposed to pure theory.

*Ethology and Sociobiology* (Elsevier North Holland, New York). Includes articles on both animal and human behavior, some of which are theoretical or philosophical in orientation.

*Journal of Social and Biological Structures* (Academic Press, New York). Devoted primarily to factual and theoretical articles on human sociobiology and its many implications in philosophy and social theory.

*Sociobiology* (California State University, Chico). Despite its encompassing title, the journal is limited to very technical articles on ants, termites, and other social insects.

*Insectes Sociaux* (Masson, Paris). The principal international journal specializing on social insects.

Among the many books on sociobiology that have appeared, both favorable and critical, are the following:

*Sociobiology: The New Synthesis*, by Edward O. Wilson (Belknap Press of Harvard University Press, Cambridge, 1975). This large book defines the field in modern terms, combining it systematically with population biology, and reviews the facts of social behavior.

*Behavior and Sociobiology*, 2nd ed., by David P. Barash (Elsevier North Holland, New York, 1981). A textbook on the social behavior of animals and human beings written for beginning students.

*Human Sociobiology: A Holistic Approach*, by Daniel G. Freedman

(Free Press, New York, 1979). A textbook that concentrates on topics in human social behavior, with special reference to psychological development.

*On Human Nature*, by Edward O. Wilson (Harvard University Press, Cambridge, 1978). Written for a broad audience, this book explores ethical and philosophical issues of human sociobiology.

*Darwinism and Human Affairs*, by Richard D. Alexander (University of Washington Press, Seattle, 1979). An extended essay comparable to *On Human Nature*, this work presents the personal views of a distinguished sociobiologist who has studied both animal and human behavior.

*Genes, Mind, and Culture*, by Charles J. Lumsden and Edward O. Wilson (Harvard University Press, Cambridge, 1981). The full monographic presentation of the theory of gene-culture coevolution and attendant evidence.

*The Woman That Never Evolved*, by Sarah Blaffer Hrdy (Harvard University Press, Cambridge, 1981). A highly acclaimed account of the evolution of female behavior in primates, with special reference to the origin of human sexuality and sex differences. It is sometimes said to add a feminist perspective to sociobiology, but it goes much further and is an important scholarly work independent of any ideology.

*Evolutionary Biology and Human Social Behavior: An Anthropological Perspective*, edited by Napoleon A. Chagnon and William Irons (Duxbury Press, North Scituate, 1979). State-of-the-art articles and essays in human sociobiology and related topics of cultural anthropology.

*Sex, Evolution and Behavior: Adaptations for Reproduction*, by Martin Daly and Margo Wilson (Duxbury Press, North Scituate, 1978). A scholarly textbook on the origin and regulation of human sexuality, incorporating both physiology and sociobiology.

*The Evolution of Human Sexuality*, by Donald Symons (Oxford University Press, New York, 1979). An excellent but orthodox view of the subject; should be read in conjunction with Daly and Wilson's *Sex, Evolution and Behavior* and Hrdy's *The Woman That Never Evolved*.

*Evolution of Social Behavior: Hypotheses and Empirical Tests*, edited by Hubert Markl (Life Sciences Research Report 18, Dahlem Konferenzen, Verlag Chemie, Deerfield Beach, Fla., 1980). A series of essays and conference reports, often of a technical nature, by leading researchers in evolutionary theory and sociobiology.

*The Expanding Circle: Ethics and Sociobiology,* by Peter Singer (Farrar, Straus and Giroux, New York, 1981). A probing extended essay on the meaning of modern evolutionary biology for ethical philosophy.

*The Use and Abuse of Biology: An Anthropological Critique of Sociobiology,* by Marshall Sahlins (University of Michigan Press, Ann Arbor, 1976). In one of the first truly scholarly critiques of the human applications, Sahlins examines the usefulness of the theory of kin selection for ethnographic studies and finds it wanting.

*The Sociobiology Debate: Readings on Ethical and Scientific Issues,* edited by Arthur L. Caplan (Harper and Row, New York, 1978). An excellent collection of essays both for and against human applications of sociobiology.

*Sociobiology Examined,* edited by Ashley Montagu (Oxford University Press, New York, 1980). A series of essays, mostly on human sociobiology and primarily critical.

*Human Nature and History: A Response to Sociobiology,* by Kenneth Bock (Columbia University Press, New York, 1980). On the inadequacy of sociobiology in explaining history and complex social processes, by a distinguished historian.

"Lumsden-Wilson Theory of Gene-Culture Coevolution," by Joseph S. Alper and Robert V. Lange, *Proceedings of the National Academy of Sciences,* 27(6): 3976–3979 (1981). A technical critique of the authors' model that translates individual cognition into patterns of cultural diversity. Alper and Lange question the simplicity of some of the assumptions, and they go much further—by expressing doubt that any mathematical treatment can ever be invented that successfully connects biology and the social sciences.

"Open Peer Commentary," by twenty-three authors in *Behavioral and Brain Sciences,* 5: 1–37 (1982), provides a diversity of opinions on the fundamental theory of gene-culture coevolution presented in *Genes, Mind, and Culture* by Lumsden and Wilson. The participants include geneticists, psychologists, anthropologists, sociologists, and philosophers. Together they closely examine most aspects of the theory and its supporting evidence. Although no clear consensus is reached, some difficulties are identified and some important ideas for new directions in sociobiological research are suggested.

39 Jonathan Beckwith and fourteen cosigners, "Against Sociobiology," *New York Review of Books,* November 13, 1975.

39   Ruth Hubbard was quoted in *The Harvard Crimson*, Cambridge, November 3, 1975.

40   The sources of Nazi and Soviet pseudo-genetics are described by Loren R. Graham in *Between Science and Values* (Columbia University Press, New York, 1981). Graham also presents a balanced account of the history and political implications of sociobiology.

41   The primary technical articles on the heritability of traits of human behavior are cited by Lee Ehrman and Peter A. Parsons, *Behavior, Genetics, and Evolution* (McGraw-Hill, New York, 1981), and Lumsden and Wilson, *Genes, Mind, and Culture.*

43   Jonathan Beckwith, "The Struggle at Harvard," *Science for the People*, 9: 31 (March–April 1977). The genetics project halted at the Harvard Medical School was the study of the development of XYY males; the directors of the project were Stanley Walzer and Park Gerald. An account of the principal events, including the involvement of Science for the People, is given by Barbara J. Culliton, "XYY: Harvard Researcher Under Fire Stops Newborn Screening," *Science*, 188: 1284–1285 (1975); and Bernard D. Davis, "XYY: The Dangers of Regulating Research by Adverse Publicity," *Harvard Magazine*, 79: 26–30 (October 1976).

43   Tom Bethell, "Burning Darwin To Save Marx," *Harper's*, December 1978, pp. 31–38, 91–92.

44   Scholarly books that take a generally favorable stance toward human sociobiology and its role in the social sciences and humanities include Mary Midgley, *Beast and Man* (Cornell University Press, Ithaca, 1978); Michael Ruse, *Sociobiology: Sense or Nonsense?* (Reidel, Boston, 1979), and *Is Science Sexist?* (Reidel, Boston, 1981); Alexander Rosenberg, *Sociobiology and the Preemption of Social Science* (Johns Hopkins University Press, Baltimore, 1980); Peter Singer, *The Expanding Circle: Ethics and Sociobiology* (Farrar, Straus and Giroux, New York, 1981); Donald Symons, *The Evolution of Human Sexuality* (Oxford University Press, New York, 1979); and Pierre L. van den Berghe, *Human Family Systems: An Evolutionary View* (Elsevier North Holland, New York, 1979). The most critical works include Bock, *Human Nature and History*; Marvin Harris, *Cultural Materialism: The Struggle for a Science of Culture* (Random House, New York, 1979); Montagu, ed., *Sociobiology Examined*; and Sahlins, *The Use and Abuse of Biology.*

44   Stuart Hampshire, "The Illusion of Sociobiology," *New York Review of Books*, October 12, 1978.

46  The content and epistemological implications of quantum theory are set out in two works by Max Jammer, *The Conceptual Development of Quantum Mechanics* (McGraw-Hill, New York, 1966), and *The Philosophy of Quantum Mechanics* (Wiley, New York, 1974). Past and recent controversies have recently been summarized by Stephen Brush, "The Chimerical Cat: Philosophy of Quantum Mechanics in Historical Perspective," *Social Studies of Science*, 10: 393–447 (1980), and by B. d'Espagnat, "The Quantum Theory and Reality," *Scientific American*, 241(5): 158–181 (November 1979).

47  For various aspects of the drive toward autonomy in the social sciences, see the historical reviews by Marvin Harris, *The Rise of Anthropological Theory* (Crowell, New York, 1968); Elvin Hatch, *Theories of Man and Culture* (Columbia University Press, New York, 1973); Annemarie de Waal Malefijt, *Images of Man: A History of Anthropological Thought* (Knopf, New York, 1974); and Frank J. Sulloway, *Freud: Biologist of the Mind* (Basic Books, New York, 1979).

48  The history of the study of gene-culture coevolution is given in Lumsden and Wilson, *Genes, Mind, and Culture.*

49  Gerald Holton, "Where Is Science Taking Us?", Jefferson Lecture in the Humanities, 1981, quoted by Zöe Ingalls in *Chronicle of Higher Education*, May 18, 1981, pp. 3–4.

**The Rules of Mental Development**

53  Konrad Lorenz, *Behind the Mirror* (Harcourt Brace Jovanovich, New York, 1977).

53  The existence of intelligent life on other planets is of course a subject of intense dispute among scientists. A few, such as Philip Morrison and Carl Sagan, have argued that 100,000 or more technical civilizations might exist in the Milky Way galaxy. Others, including Robert T. Rood and James S. Trefil in *Are We Alone?* (Scribner's, New York, 1981), conclude that intelligent life is highly improbable. We invented the civilizations of the eidylons and xenidrins to provide a livelier and clearer background for the theory of the origin of human intelligence.

54  The directed learning of bird song is reviewed by John Alcock in *Animal Behavior*, 2nd ed. (Sinauer Associates, Sunderland, Mass., 1979).

55    DNA, the basic hereditary material, does not appear adequate to create under terrestrial conditions a fully genetic civilization such as that ascribed to the distant eidylons. There are several hundred thousand human genes made up of about 2.9 billion nucleotide pairs, the basic chemical units of DNA. A change in any one of the nucleotide pairs is by definition a mutation. It alters the gene of which it is a part and can change one or more traits in anatomy, physiology, or behavior. The number of nucleotides found in the human cell is not very far below the maximum possible. Some plants, salamanders, and fishes have managed to accumulate between 10 and 100 million pairs during evolution (without, incidentally, becoming any bigger, more complicated, or smarter). But it would be difficult or impossible to fit much more DNA than that into individual cells of ordinary size. The DNA and its supporting protein matrix would become so bulky as to create severe problems in nourishment and transport. In short, the fundamental properties of life as we know it limit the capacity of organisms to store information within their own bodies.

So what can be done with a billion or so nucleotide pairs? If there were some way to translate nucleotides directly into the nodes of long-term memory, such as the concepts for *white, horse,* and *running,* no less than a billion such pieces of information could be programed into the brain. Suppose that each node were also linked to ten other modes to form a more complex concept, for example *white* to *horse* to *running.* The nucleotides, if permitted to transfer their information in a direct manner, could still program as many concepts as the human brain actually learns by experience. On the other hand, the powers of the human genes would soon be exhausted if they were required to encode language. Although a very large number of symbols can be innately programed and linked to other nodes to label concepts, there is a strict limit to the length of sentences that can be created in an instinctual manner. To possess a completely inborn vocabulary of 10,000 words and to speak in completely inborn sentences of ten words each would require an astronomical $10^{40}$ nucleotides (a 1 followed by 40 zeroes), or one hundred million billion kilograms of DNA—far more than the weight of the entire human species.

55    This information analysis of genes, neurons, and messages was first undertaken by H. J. Bremermann in "Limits of Genetic Control," *IEEE Transactions on Military Electronics,* MIL-7 (2 and 3):

200–205 (1963). It was considerably expanded and refined by us in *Genes, Mind, and Culture*.

55   Our reasoning on the ultimate capacity of human memory is as follows: Consider that there are $3.15 \times 10^7$ seconds in a year, so that during a generous life span of 100 years, $3.15 \times 10^9$ seconds at most are available to acquire new information. At a minimum of 10 seconds required to acquire each new concept in long-term memory (the time determined by experiments), individuals can accumulate $3.15 \times 10^8$ concepts in their lifetime if they do nothing else but learn. If 10 percent of the life span is devoted to learning, a maximum of $3.15 \times 10^7$ concepts can be assimilated. If a more modest 0.1 percent of the life span is used, the maximum will be $3.15 \times 10^5$ concepts.

56   The statements by Leslie A. White are from "Individuality and Individualism: A Culturological Interpretation," *Texas Quarterly*, 6: 111–127 (1963); and his review of *From Savagery to Civilization* by G. Clarke and *History* by V. G. Childe, published in *Antiquity*, 22: 217–218 (1948).

61   For an account of !Kung socialization, see Patricia Draper's "Social and Economic Constraints on Child Life among the !Kung," in R. B. Lee and Irven DeVore, eds., *Kalahari Hunter-Gatherers: Studies of the !Kung San and Their Neighbors* (Harvard University Press, Cambridge, 1976). The information on the Ituri pygmies and other preliterate people comes from Irven DeVore (personal communication). For a general cross-cultural analysis see Herbert Barry, III, et al., "Relation of Child Training to Subsistence Economy," *American Anthropologist*, 61: 51–63 (1959).

64   The twelve categories we found on which quantitative choice studies had been conducted are: dietary preference in infants, proportionate representation of vision and the other principal senses in vocabularies, color classification, phoneme formation in the development of language, infant preference for certain geometric designs, infant preference for normally composed facial features, the choice of facial expressions to convey basic emotions, the forms of mother-infant bonding and communication, the mode of carrying infants and intermediate-sized objects, phobias, incest avoidance, and reaction to strangers. These studies are reviewed in detail in *Genes, Mind, and Culture*.

64   The brother-sister incest case is discussed further, with documentation, in the final two chapters of this book.

65   The basic research on infant color perception is reported by Marc H. Bornstein, "Perceptual Development: Stability and Change in Feature Perception," in M. H. Bornstein and W. Kessen, eds., *Psychological Development from Infancy: Image to Intention* (Lawrence Erlbaum, Hillsdale, 1979), pp. 37–81.

65   The experiments on color vocabulary are cited in Brent Berlin and Paul Kay, *Basic Color Terms: Their Universality and Evolution* (University of California Press, Berkeley, 1969).

65   The determinants of color vision are described by George Wald, "The Molecular Basis of Color Vision," in B. R. Straatsma et al., eds., *The Retina: Morphology, Function, and Clinical Characteristics* (University of California Press, Berkeley, 1969), pp. 281–295; Peter H. Lindsay and Donald A. Norman, in *Human Information Processing: An Introduction to Psychology*, 2nd ed. (Academic Press, New York, 1977).

66   The genetics of color blindness is reviewed by Curt Stern, *Principles of Human Genetics*, 3rd ed. (Freeman, San Francisco, 1973).

67   Learning of color vocabularies by the Dani of New Guinea: Eleanor H. Rosch, "Natural Categories," *Cognitive Psychology*, 4: 328–350 (1973).

67   The preference of newborn infants for certain geometric designs is documented by R. L. Fantz, J. F. Fagan, III, and S. B. Miranda, "Early Visual Selectivity: As a Function of Pattern Variables, Previous Exposure, Age from Birth and Conception, and Expected Cognitive Deficit," in L. B. Cohen and P. Salapatek, eds., *Infant Perception, I: Basic Visual Processes* (Academic Press, New York, 1975), pp. 249–345. The optimum number of turns in figures was demonstrated by M. Hershenson, H. Munsinger, and W. Kessen in "Preference for Shapes of Intermediate Variability in the Newborn Human," *Science*, 147: 630–631 (1965).

68   The amount of brain arousal caused by geometric designs of differing complexity was studied by Gerda Smets, a Belgian psychologist, and reported in *Aesthetic Judgment and Arousal: An Experimental Contribution to Psycho-Aesthetics* (Leuven University Press, Leuven, Belgium, 1973). Complexity was measured by the percentage of redundancy in the individual spots that composed the figure. The arousal measure was the duration of the period during which the alpha wave of the electroencephalogram was blocked after the presentation of the figure. This purely physiological response is considered to be linked to the ultimate es-

thetic judgment of such designs and the preference for one over another in art and symbolism.

68 The experiments on basic emotional expressions are reported by Paul Ekman in *The Face of Man: Expressions of Universal Emotions in a New Guinea Village* (Garland STPM Press, New York, 1980).

69 The taste preferences of infants and young children are documented by O. Maller and J. A. Desor, "Effects of Taste on Ingestion by Human Newborns," in J. Bosma, ed., *Fourth Symposium on Oral Sensation and Perception: Development in the Fetus and Infant* (Government Printing Office, Washington, D.C., 1974), pp. 279–311; M. Chiva, "Comment la personne se construit en mangeant," *Communications* (École des Hautes en Sciences Sociales—Centre d'Études Transdisciplinaires, Paris), 31: 107–118 (1979); and J. E. Steiner, "Oral and Facial Innate Motor Responses to Gustatory and to Some Olfactory Stimuli," in J. H. A. Kroeze, ed., *Preference Behaviour and Chemoreception* (Information Retrieval, London, 1979), pp. 247–261.

69 The fear response of children in the presence of strangers is described by Irenäus Eibl-Eibesfeldt in "Human Ethology: Concepts and Implications for the Sciences of Man," *Behavioral and Brain Sciences*, 2: 1–57 (1979).

70 The role of the eyes in communication is discussed in M. Argyle and M. Cook, *Gaze and Mutual Gaze* (Cambridge University Press, Cambridge, 1976). New evidence of the deep biological origin of the threat stare is presented by Kent G. Bailey et al. in "The Threatening Stare: Differential Response Latencies in Mild and Profoundly Retarded Adults," *American Journal of Mental Deficiency*, 81: 599–602 (1977).

71 The classification of people by stereotype is documented by Amos Tversky and Daniel Kahneman, "Judgment under Uncertainty: Heuristics and Biases," *Science*, 185: 1124–1131 (1974).

71 Risk assessment by human beings is evaluated by L. Reijnders, "On the Applicability of Game Theory to Evolution," *Journal of Theoretical Biology*, 75: 245–247 (1978); and D. W. Orr, "Catastrophe and Social Order," *Human Ecology*, 7: 41–52 (1979).

73 Accounts of the genetic mutations that affect the cerebellar structure and locomotion of mice are given by Pasco Rakic, "Genetic and Epigenetic Determinants of Local Neuronal Circuits in the Mammalian Central Nervous System," in F. O. Schmitt and

F. G. Worden, eds., *The Neurosciences: Fourth Study Program* (MIT Press, Cambridge, 1979), pp. 109–127; and G. E. McClearn and J. C. DeFries, *Introduction to Behavioral Genetics* (W. H. Freeman, San Francisco, 1973).

73   An excellent account of Turner syndrome, the set of conditions caused by an absence of a sex chromosome, is given by William L. Nyhan and Edward Edelson, *The Heredity Factor* (Grosset and Dunlap, New York, 1976).

73   The major gene having selective effects on various kinds of cognitive ability was reported by G. C. Ashton, J. J. Polovina, and S. G. Vandenberg, "Segregation Analysis of Family Data for 15 Tests of Cognitive Ability," *Behavior Genetics*, 9(5): 329–347 (1979).

73   The number of identified human genes is reported by Victor A. McKusick, "The Anatomy of the Human Genome," *Journal of Heredity*, 71: 370–391 (1980).

74   The simple genetic control of the ability to smell a certain form of musk has been reported by D. Whissell-Buechy and J. E. Amoore, "Odour-Blindness to Musk: Simple Recessive Inheritance," *Nature*, 242: 271–273 (1973).

74   The Lesch-Nyhan syndrome is described by one of its discoverers in Nyhan and Edelson, *The Heredity Factor*.

75   An account of facial communication is given by Paul Ekman, "Cross-cultural Studies of Facial Expression," in P. Ekman, ed., *Darwin and Facial Expression: A Century of Research in Review* (Academic Press, New York, 1973).

75   This statement by Eccles on the transcendence of the mind is quoted from the preface to *Critique of the Psycho-Physical Theory*, by E. P. Polton (Mouton, The Hague, 1973). See also Karl R. Popper and J. C. Eccles, *The Self and Its Brain* (Springer International, New York, 1977), and J. C. Eccles, *The Human Psyche* (Springer International, New York, 1980).

76   Most of the review of the mind-body problem is drawn from William R. Uttal, *The Psychobiology of Mind* (Lawrence Erlbaum, Hillsdale, 1978), and Jerry A. Fodor, "The Mind-Body Problem," *Scientific American*, 244(1): 114–123 (January 1981).

77   The research on electroencephalogram variation is reported by E. R. John et al. in "Developmental Equations for the Electroencephalogram," *Science*, 210: 1255–1258 (1980); and H. Ahn et al.,

"Developmental Equations Reflect Brain Dysfunctions," *Science*, 210: 1259–1262 (1980).

77 Recent research on PET analysis of brain activity is reported by Julie Ann Miller, "Brain-Watch," *Science News*, 119: 76–78 (1981).

77 The xenon 133 technique used to monitor blood flow in the brain during various forms of mental activity is reported by Niels A. Lassen, David H. Ingvar, and Erik Skinhøj, "Brain Function and Blood Flow," *Scientific American*, 239(4): 62–71 (October 1978).

79 Reviews of the psychology of memory are given by Geoffrey R. Loftus and Elizabeth F. Loftus, *Human Memory: The Processing of Information* (Lawrence Erlbaum Associates, Hillsdale, 1976), and P. H. Lindsay and D. A. Norman, *Human Information Processing: An Introduction to Psychology*, 2nd ed. (Academic Press, New York, 1977). Specific details of the spreading-activation theory have been questioned, but the basic conception appears consistent with a growing body of experimental evidence. Furthermore, it is our impression that the chief rival models are actually complementary to the conception of networks of node-link structures.

85 Peter C. Reynolds, *On the Evolution of Human Behavior* (University of California Press, Berkeley, 1981).

## The Social Worlds of *Homo*

86 The body form of *Homo habilis* is based on the reconstruction by J. H. Matternes, as presented for example in *National Geographic*, 143: 819–829 (June 1973).

88 The figure comparing human and chimpanzee chromosomes is based on J. J. Yunis et al., "The Striking Resemblance of High Resolution G-banded Chromosomes of Man and Chimpanzee," *Science*, 208: 1145–1148 (1980). The close similarity between human beings and chimpanzees at the level of the gene is documented by Mary-Claire King and A. C. Wilson, "Evolution at two levels in humans and chimpanzees," *Science*, 188: 107–116 (1975). Those authors note that the amino acids composing the average protein are more than 99 percent identical between the two species. On the basis of combined fossil evidence and biochemical differences in the living species (specifically, the structure of fibrinopeptides, which are small proteins), the chimpanzee and human lines are estimated to have split from each other sometime between 5 and 20 million years ago. See A. C. Wilson,

S. S. Carlson, and T. J. White, "Biochemical Evolution," *Annual Review of Biochemistry*, 46: 573–639 (1977).

89    The technique used for deducing the traits of common ancestors and intermediate species is standard in the study of evolution. See for example E. O. Wiley, *Phylogenetics: The Theory and Practice of Phylogenetic Systematics* (Wiley, New York, 1981).

90    The use of the hand in gesturing others away was observed in free-ranging chimpanzees at the Gombe Stream National Park, Tanzania; see Suzanne Chevalier-Skolnikoff, "The Gestural Abilities of Apes," *Behavioral and Brain Sciences*, 2: 382–383 (1979).

90    The description of band size and mingling in *Homo habilis* is based on the common traits of chimpanzee and human hunter-gatherer bands. See, for example, Wilson, *Sociobiology*; C. J. Buys and K. L. Larson, "Human Sympathy Groups," *Psychological Reports*, 45: 547–553 (1979); and R. B. Lee, *The !Kung San: Men, Women, and Work in a Foraging Society* (Cambridge University Press, New York, 1979).

90    The common properties of play in animals and men are described in an evolutionary context in Robert M. Fagen's definitive monograph *Animal Play Behavior* (Oxford University Press, New York, 1981).

91    The facial expressions depicted in the illustration of *Homo habilis* are among the ones that are similar between chimpanzees and modern human beings and appear to be associated with the same emotions. Using the Y technique of evolutionary reconstruction, these expressions have also been attributed to *Homo habilis*. The sources of the comparisons are as follows. Human laughter: in chimpanzees the equivalent is the "relaxed open-mouth display" with "ah-ah" vocalization; based on J.A.R.A.M. van Hooff, "A Comparative Approach to the Phylogeny of Laughter and Smiling," in R. A. Hinde, ed., *Non-Verbal Communication* (Cambridge University Press, New York, 1972). The depiction and interpretation of pouting is modified from Peter Marler and Richard Tenaza, "Signaling Behavior of Apes with Special Reference to Vocalization," in T. A. Sebeok, *How Animals Communicate* (Indiana University Press, Bloomington, 1977). The representation of anger is based on van Hooff.

92    The hunting (or scavenging) of a hippopotamus by *Homo habilis* is based loosely on an archaeological reconstruction reported by Glynn Isaac in "The Food-Sharing Behavior of Protohuman Hominids," *Scientific American*, 238(4): 90–108 (April 1978). Isaac

is also among recent authors who have come to stress the importance of a home base and food sharing in the origins of man. The details of the butchery are taken from a general account by J. Wallace, "Evolutionary Trends in the Early Hominid Dentition," in Clifford J. Jolly, ed., *Early Hominids of Africa* (St. Martin's Press, New York, 1978), pp. 285–310.

92  The probable number of blows required to produce a stone chopper of the kind used by *Homo habilis* is based on Sergei A. Semenov, *Prehistoric Technology: An Experimental Study of the Oldest Tools and Artefacts*, trans. M. W. Thompson (Adams and Dart, Bath, Somerset, 1964). See the further discussion by Ralph L. Holloway, "Culture, Symbols, and Human Brain Evolution," *Dialectical Anthropology*, 5: 287–303 (1981).

92  The leading role of males in hunting and meat sharing, as well as the intense excitement during the sharing, occurs in both chimpanzees and modern hunter-gatherers. See, for example, Geza Teleki, *The Predatory Behavior of Wild Chimpanzees* (Bucknell University Press, Lewisberg, 1973); and John E. Pfeiffer, *The Emergence of Man* (Harper and Row, New York, 1969).

92  The reconstruction of the ecology and habits of *Homo habilis* are reviewed by Karl W. Butzer, "Environment, Culture, and Human Evolution," *American Scientist*, 65: 572–584 (1977); Glynn Isaac, "Casting the Net Wide: A Review of Archaeological Evidence for Early Hominid Land-Use and Ecological Relations," in L.-K. Königsson, ed., *Current Arguments on Early Man* (Pergamon Press, New York, 1980), pp. 114–134; and Anna K. Behrensmeyer, Butzer, Isaac, C. O. Lovejoy, et al., in *Early Hominids of Africa*, Clifford J. Jolly, ed. (St. Martin's, New York, 1978).

94  The behavioral flexibility and intellectual capabilities of chimpanzees and other apes are examined in excellent reviews by Gordon G. Gallup et al., "A Mirror for the Mind of Man, or Will the Chimpanzee Create an Identity Crisis for *Homo sapiens*?", *Journal of Human Evolution*, 6: 30–313 (1977), and Sue Taylor Parker and Kathleen Rita Gibson, "A Developmental Model for the Evolution of Language and Intelligence in Early Hominids," *Behavioral and Brain Sciences*, 2: 367–408 (1979). In his *Foundations of Primitive Thought* (Oxford University Press, New York, 1979), C. R. Hallpike provides a thoroughgoing review of research on cognition in hunter-gatherer and other primitive societies of modern man, examined primarily from a Piagetian point of view.

94  The importance of play in the discovery of new, adaptive forms

of behavior is stressed by Robert M. Fagen, *Animal Play Behavior* (Oxford University Press, New York, 1981), and Brian Vandenberg, "The Role of Play in the Development of Insightful Tool-Using Strategies," *Merrill-Palmer Quarterly*, 27: 97–109 (1981).

94    The story of how the squirrel monkey Corwin invented play routines is told by John D. and Janice I. Baldwin, "The Role of Learning Phenomena in the Ontogeny of Exploration and Play," in Suzanne Chevalier-Skolnikoff and Frank E. Poirier, eds., *Primate Bio-Social Development* (Garland, New York, 1977), pp. 343–406.

95    The use of kerosene-can banging in the threat display of a male chimpanzee is recounted by Jane van Lawick-Goodall, *In the Shadow of Man* (Houghton Mifflin, Boston, 1971).

95    The story of the crippled chimpanzee is given by Geza Teleki in *The Predatory Behavior of Wild Chimpanzees*.

95    The Piagetian analysis of behavior development in monkeys and apes is reviewed by Parker and Gibson, "A Developmental Model," as well as by Suzanne Chevalier-Skolnikoff in "A Piagetian Model for Describing and Comparing Socialization in Monkey, Ape, and Human Infants," in Chevalier-Skolnikoff and Poirier, eds., *Primate Bio-Social Development*, pp. 159–187. The artistic endeavors of captive chimpanzees are nicely exhibited in Desmond Morris, *The Biology of Art* (Knopf, New York, 1962).

96    The development of incest avoidance by chimpanzees is documented by Anne E. Pusey, "The Physical and Social Development of Wild Adolescent Chimpanzees (*Pan troglodytes schweinfurthii*)," Ph.D. diss., Stanford University (1977); cited by Craig Packer in *Animal Behaviour*, 27: 1–36 (1979). Pusey found that individuals who grew up in close association (and hence were most likely to be siblings) were least likely to attempt copulation at maturity. Incest is further reduced by the tendency of young females to migrate to neighboring groups.

96    The extreme "phobic" reaction of chimpanzees to a stuffed leopard was reported by A. Kortlandt and M. Kooij, "Protohominid Behaviour in Primates (Preliminary Communication)," *Symposia of the Zoological Society of London*, 10: 61–88 (1963).

97    Various aspects of the reconstruction of the social life and ecology of *Homo erectus* are provided by Karl W. Butzer, "Environment, Culture, and Human Evolution," *American Scientist*, 65: 579

(1977); W. W. Howells, *"Homo erectus*—Who, When, and Where: A Survey," *Yearbook of Physical Anthropology,* 23: 1–23 (1980); and Jan Jelínek, "European *Homo erectus* and the Origin of *Homo sapiens,*" in Königsson, ed., *Current Arguments on Early Man.* F. Clark Howell's account of the possible use of fire by *Homo erectus* is quoted by Pfeiffer in *The Emergence of Man.* The date of the earliest use of fire at campsites is examined in "Early Hominids and Fire at Chesowanja, Kenya," an exchange between Glynn Isaac and J.A.J. Gowlett et al. in *Nature,* 296: 870 (1982). The exchange of young females between bands is a trait shared by chimpanzees and some modern hunter-gatherer societies; see Anne E. Pusey, "Intercommunity transfer of chimpanzees ir Gombe National Park," in D. A. Hamburg and Elizabeth R. McCown, *The Great Apes* (Benjamin/Cummings, Menlo Park, 1979), pp. 465–479. The same is true of fatal battles between males. Our account of a primitive symbolic language by *Homo erectus* is pure guesswork.

98   The *Homo erectus* figure is based on various authorities, including especially Zdenek Burian in *Science Digest,* 89: 41 (1981), and F. C. Howell, *Early Man* (Time Inc., New York, 1965).

101   The slow progress and occasional retreat of *Homo erectus* culture is described by Butzer, "Environment, Culture, and Human Evolution," *American Scientist,* 65: 579 (1977).

101   The argument that social arrangements were primary and the materials culture secondary in the mental evolution of *Homo* was advanced by Ralph L. Holloway in "Culture, Symbols, and Human Brain Evolution: A Synthesis," *Dialectical Anthropology,* 5: 287–303 (1981).

102   The diagram of the evolution of the hominid brain is based on Ralph L. Holloway, "The Casts of Fossil Hominid Brains," *Scientific American,* 231(1): 106–115 (July 1974); Edgar B. Zurif, "Language Mechanisms: A Neuropsychological Perspective," *American Scientist,* 68: 305–311, 1980; and R. E. Passingham, *The Human Primate* (Freeman, San Francisco, 1982).

103   A fuller account of the decipherment of Paleolithic "writing" is given by Alexander Marshack, "Upper Paleolithic Symbol Systems of the Russian Plain: Cognitive and Comparative Analysis," *Current Anthropology,* 20: 271–311 (1979); and Clive Gamble, "Information Exchange in the Palaeolithic," *Nature,* 283: 522–523 (1980).

104   The figure of the Tallensi compound is adapted from Meyer

Fortes, "Primitive Kinship," *Scientific American,* 200(6): 146 (June 1959).

105   The evidence that symbolic thought preceded true verbal language is based on experiments indicating that chimpanzees and other apes can transfer concepts from one sensory modality (such as touch) to another (such as vision). See A. J. Premack and David Premack, "Teaching Language to an Ape," *Scientific American,* 227(4): 92–99 (October 1972); G. Ettlinger, "The Transfer of Information Between Sense-Modalities: A Neuropsychological Review," in H. P. Zippel, ed., *Memory Transfer and Information* (Plenum, New York, 1973); and G. H. Hewes, "Primate Communication and the Gestural Origin of Language," *Current Anthropology,* 14: 5–12 (1973).

105   Although many competent scientists believe on the basis of substantial evidence that chimpanzees can learn symbolic language to the extent of inventing new phrases and sentences to express elementary concepts, others are convinced that the responses obtained during language training consist of rote learning no different from that of rats mastering laboratory mazes. Both sides of the argument were recently examined in a volume edited by Thomas A. Sebeok and Robert Rosenthal, *The Clever Hans Phenomenon: Communication with Horses, Whales, Apes, and People* (New York Academy of Sciences, New York, 1981). The evidence appears to us to be consistent with a limited capacity to generate language under human guidance. It does not seem outrageous to suppose that on a scale of 0 to 100 in language ability, with rats 0 and humans 100, chimpanzees might fall somewhere between 0.1 and 1.

105   Gestural babbling by infant chimpanzees is described by Gordon G. Gallup et al., "A Mirror for the Mind of Man, or Will the Chimpanzee Create an Identity Crisis for *Homo sapiens*?", *Journal of Human Evolution,* 6: 303–313 (1977).

106   The figure of Paleolithic *Homo sapiens* is based on original specimens and various published sources, including Josef Wolf's *The Dawn of Man* (Abrams, New York, 1978).

107   Recent findings on the size and structure of the brain and the relation of brain anatomy to intelligence and language capacity are presented by the multiple authors of *Development and Evolution of Brain Size: Behavioral Implications* (Academic Press, New York, 1979), as well as by Edgar B. Zurif in "Language Mechanisms: A

Neuropsychological Perspective," *American Scientist*, 68: 305–311 (1980).

109   The expression "sapientization" was used by R. Parenti "to indicate all the transformations and events leading from the first form recognizable as *Homo* . . . to the emergence and establishment of the morphological, functional and psychological features which characterize the species *Homo sapiens (sapiens)*." *Journal of Human Evolution*, 2:499–508 (1973).

109   The anatomist J. Wallace has spoken of the origin of *Homo* in the following terms: "For at least 10 million years hominids had relied on their teeth for 'chewing.' At some time between 2.5 and 3 million years ago, a dentally 'gracile' australopithecine, perhaps much like those at Sterkfontein and Makapansgat, discovered a better, a more efficient way of 'chewing,' namely stones. This hominid who substituted sharpened stones for his teeth to split apart the tough outer coverings of food was the first *Homo*, the first hominid to escape from selection for reduction of cusp height and early fusion of the premaxilla. Now largely removed from teeth, selection focused on the brain, and as the record attests the brain began to blossom." In C. J. Jolly, ed., *Early Hominids in Africa* (St. Martin's Press, New York, 1978), p. 306.

109   This expansion of human and man-ape classification, which is advanced for purposes of illustration and not as a serious proposal, is nevertheless based on correct taxonomic procedures; see the existing available formal arrangement reviewed by Bernard Campbell in "The Nomenclature of the Hominidae, Including a Definitive List of Hominid Taxa," Royal Anthropological Institute of Great Britain and Ireland, Occasional Paper 22 (1965).

110   Mary LeCron Foster has presented her hypothesis of the ritualistic origin of language in "The Symbolic Structure of Primordial Language," in Sherwood Washburn and E. R. McCown, eds., *Perspectives in Human Language*, vol. 4 (Benjamin/Cummings, Menlo Park, 1978); and "The Growth of Symbolism in Culture," in M. LeCron Foster and Stanley H. Brandes, eds., *Symbol as Sense* (Academic Press, New York, 1980).

112   The epigenetic rules of language development are discussed in Noam Chomsky, *Rules and Representations* (Columbia University Press, New York, 1980), and in Kenneth Wexler and Peter W. Culicover, *Formal Principles of Language Acquisition* (MIT Press, Cambridge, 1980).

113    The M-constraint was first proposed by Fred Sommers in "The Ordinary Language Tree," *Mind*, 68: 160–185 (1959). The approach taken by experimental psychologists to this and other peculiarities in the thought process is ably reviewed by Frank C. Keil in *Semantic and Conceptual Development: An Ontological Perspective* (Harvard University Press, Cambridge, 1979), and "Constraints on Knowledge and Cognitive Development," *Psychological Review*, 88: 197–227 (1981).

113    The psychological constraints on quantitative reasoning have been examined by Rochel Gelman and C. R. Gallistel in *The Child's Understanding of Number* (Harvard University Press, Cambridge, 1978), and R. Gelman in "What Young Children Know about Numbers," *Educational Psychologist*, 15: 54–68 (1980).

## Promethean Fire

117    In the strict usage of biology, the word *coevolution* means a genetic change of one species in response to the evolution of a second species, which in turn changes in response to the first species. Thus the system of two species evolves to some extent as a unit. Coevolution is more strongly developed when the two species are mutually dependent. Bees, for example, have evolved an automatic response to certain colors and smells of flowers; they visit the flowers and are rewarded with nectar and pollen. The flowering plants in turn have developed the colors and smells by which they attract the bees; their reward is cross-pollination and successful reproduction. We stretched the term to include the reciprocal effects of genetic and cultural change within the human species. Our usage is also different from that of William Durham, "The Coevolution of Human Biology and Culture," in N. Blurton Jones and V. Reynolds, eds., *Human Adaptation and Behavior* (Halsted Press, Wiley, New York, 1978). By coevolution Durham means parallel but unlinked changes in genes and culture.

120    Friedrich Engels on the definition of history: *Ludwig Feuerbach and the Outcome of Classical German Philosophy* (International Publishers, New York, 1941), p. 49.

121    We give a full discussion of the basic unit of culture, the culturgen, in *Genes, Mind, and Culture*. We also describe the statistical techniques that can be used to define the limits of more diffuse clusters of artifacts and behaviors. Critiques and defenses of the culturgen concept are given by many of the twenty-three authors

who reviewed *Genes, Mind, and Culture* in *Behavioral and Brain Sciences*, 5: 1–37 (1982).

121   The principal psychological method by which culturgens might be matched with node-link structures in long-term memory is the semantic differential technique. An elementary account is given in G. Lindzey, C. S. Hall, and R. F. Thompson, *Psychology* (Worth, New York, 1975).

122   The account of the Tapirapé is based on Charles Wagley's *Welcome of Tears: The Tapirapé Indians of Central Brazil* (Oxford University Press, New York, 1977). Wagley's study was conducted in 1939–40, when the tribes of central Brazil were still close to their original state. Since then Tapirapé customs have been profoundly altered by contact with white men. The transformation is a tragedy for all. As Wagley says, "each of these small societies has a view of the world in its own terms and each has much to offer . . . The disappearance of societies such as [the Tapirapé] would be an irreplaceable loss to the world."

125   The details of Tapirapé sexual behavior in this fictional episode are all based on Charles Wagley's *Welcome of Tears*, except for brother-sister incest, which is based loosely on a later interview with Wagley.

126   Preferences of alternative cultural choices and rates of switching between them can be estimated by appropriately planned informant analysis. See, for example, S. C. Dodd, "Diffusion Is Predictable: Testing Probability Models for Laws of Interaction," *American Sociological Review*, 20: 392–401 (1955); and Howard Rachlin, *Behavior and Learning* (Freeman, San Francisco, 1976).

131   Solomon E. Asch, "Effects of Group Pressure upon the Modification and Distortion of Judgments," in H. Guetzkow, ed., *Groups, Leadership, and Men* (Carnegie Institute of Technology Press, Pittsburgh, 1951), pp. 170–190.

132   S. Milgram, L. Bickman, and L. Berkowitz, "Note on the Drawing Power of Crowds of Different Size," *Journal of Personality and Social Psychology*, 13: 79–82 (1969).

135   The diagram of the pattern of brother-sister incest is adapted from Figure 4-27 in our *Genes, Mind, and Culture*, p. 153.

136   Details of the injurious genetic effects of inbreeding are given by L. L. Cavalli-Sforza and W. F. Bodmer, *The Genetics of Human Populations* (Freeman, San Francisco, 1971); Eva Seemanová, "A Study

of Children of Incestuous Marriages," *Human Heredity*, 21: 108–178 (1971); and Curt Stern, *Principles of Human Genetics*, 3rd ed. (Freeman, San Francisco, 1973).

136    Information on the frequency of incest is given in G. P. Murdock, *Social Structure* (Macmillan, New York, 1949); B. Berelson and G. A. Steiner, *Human Behavior: An Inventory of Scientific Findings* (Harcourt, Brace, and World, New York, 1964); S. K. Weinberg, *Incest Behavior*, rev. ed. (Citadel Press, New York, 1976); and P. L. van den Berghe and G. M. Mesher, "Royal Incest and Inclusive Fitness," *American Ethnologist*, 7: 300–317 (1980).

137    Many social theorists have cited cultural diversity as evidence of the lack of biological influence. Among the most prominent are Marshall Sahlins, *The Use and Abuse of Biology* (University of Michigan Press, Ann Arbor, 1976); and Marvin Harris, *Cultural Materialism: The Struggle for a Science of Culture* (Random House, New York, 1979).

139    The details of Yanomamö social life were taken from Napoleon A. Chagnon, "Fission in an Amazonian Tribe," *The Sciences*, 16: 14–18 (1976), and *Yanomamö: The Fierce People*, 2nd ed. (Holt, Rinehart and Winston, New York, 1977); and from James V. Neel, "On Being Headman," *Perspectives in Biology and Medicine*, Winter 1980, pp. 277–294.

147    The basic theory of genetic evolution within specific cultural environments was developed in our *Genes, Mind, and Culture*.

149    A review of territorial aggression among hunter-gatherer bands has been provided by Glenn E. King, "Society and Territory in Human Evolution," *Journal of Human Evolution*, 5: 323–332 (1976).

156–157    The dinosaur figures are adapted from Dale A. Russell, for example in "The Mass Extinction of the Late Mesozoic," *Scientific American*, 246(1): 65 (January, 1982).

158    The olfactory system of the silkworm moth is described by Dietrich Schneider, "Insect Olfaction: Deciphering System for Chemical Messages," *Science*, 163: 1031–1037 (1969).

159    Communication by odor in domestic cats was analyzed by Paul Leyhausen, "The Communal Organization of Solitary Mammals," *Symposia of the Zoological Society of London*, 14: 249–263 (1965).

160    Seymour S. Kety, "Disorders of the Human Brain," *Scientific American*, 241(3): 202–214 (September 1979).

160 The prime-mover hypotheses are reviewed by John E. Pfeiffer, *The Emergence of Man* (Harper and Row, New York, 1969); Wilson, *Sociobiology*; and Dean Falk, "Hominid Brain Evolution: The Approach from Paleoneurology," *Yearbook of Physical Anthropology,* 23:93–107 (1980).

161 Robin Fox, "The Cultural Animal," in J. F. Eisenberg and W. S. Dillon, eds., *Man and Beast: Comparative Social Behavior* (Smithsonian Institution Press, Washington, D.C., 1971), pp. 273–296.

161 Sarah Blaffer Hrdy, *The Woman That Never Evolved* (Harvard University Press, Cambridge, 1981).

161 Charles Darwin, *The Descent of Man, and Selection in Relation to Sex,* 2 vols. (Appleton, New York, 1871).

162 Valerius Geist, *Life Strategies, Human Evolution, Environmental Design: Toward a Biological Theory of Health* (Springer-Verlag, New York, 1978).

164 A history of the Promethean myth is given in Michael Grant, *Myths of the Greeks and Romans* (Weidenfeld and Nicolson, London, 1962).

165 The two sentences from Aeschylus come from *Prometheus Bound* as translated by David Grene in *Aeschylus II,* edited by David Grene and Richard Lattimore as part of *The Complete Greek Tragedies* (University of Chicago Press, Chicago, 1956).

## Toward a New Human Science

168 The behaviorist philosophy, as conceived by its principal modern architect B. F. Skinner (see for example *About Behaviorism,* Knopf, New York, 1974), is a severe form of scientific reductionism, which rejects conceptions of mental phenomena that cannot be tested experimentally. The history of cognitive psychology, and the manner in which proponents of the new discipline relaxed this restriction to explore the mind, is reviewed by John R. Anderson, *Cognitive Psychology and Its Implications* (W. H. Freeman, San Francisco, 1980).

168 We have summarized the rules of mental development in *Genes, Mind, and Culture,* especially in Chapters 2 and 3.

170 The relation of cognitive psychology and sociobiology is discussed at greater length in Chapter 8 of *Genes, Mind, and Culture.*

172 Critiques of the biological approach to the study of the social sci-

ences and history are given in Kenneth Bock, *Human Nature and History: A Response to Sociobiology* (Columbia University Press, New York, 1980), and Ashley Montagu, ed., *Sociobiology Examined* (Oxford University Press, New York, 1980).

173    An excellent discussion of the relation between free will and intention is presented by Joseph F. Rychlak in "Concepts of Free Will in Modern Psychological Science," *Journal of Mind and Behavior*, 1(1): 9–32 (1980).

175    The role of evolutionary theory in ethical philosophy is discussed by Wilson, *On Human Nature*, and Peter Singer, *The Expanding Circle: Ethics and Sociobiology* (Farrar, Straus and Giroux, New York, 1981).

175    The suggestion that incest taboos are culturally determined and can be abolished has been made, for example, by Yehudi Cohen in "The Disappearance of the Incest Taboo," *Human Nature*, 1(7): 72–78 (July 1978).

177    Phenylketonuria and other genetic diseases are described in a simple but authoritative manner by William L. Nyhan and Edward Edelson, *The Heredity Factor* (Grosset and Dunlap, New York, 1976).

179    An additional analysis of the evolution of ethical behavior as a neurobiological phenomenon has been provided by George E. Pugh, *The Biological Origin of Human Values* (Basic Books, New York, 1977).

179    In *The Red Lamp of Incest* (Dutton, New York, 1980) Robin Fox brilliantly relates the Freudian primal-horde theory of the incest taboo to modern sociobiology.

182    The writers who have most explicitly discussed the possible existence of ethical values external to those revealed by biological analysis include Peter Singer, *The Expanding Circle;* and Robert Nozick, *Philosophical Explanations* (Belknap Press of Harvard University Press, Cambridge, 1981).

# Index

Chiva, M., 197
Chomsky, Noam, 112, 205
cities, 16
civilization, 16
Clarke, G., 195
coevolution, defined, 206
cognition, *see* color perception; epigenetic
    rules; language; memory; mind
cognitive dissonance, 5
Cohen, L. B., 196
Cohen, Yehudi, 210
colonial invertebrates, 34–35
color: perception, 65–67, 71, 115, 168; vo-
    cabulary, 65–67
communication: evolution, 109–112; sat-
    ellite, 17. *See also* language; symbolism
computers, 17, 76
Comte, Auguste, 171
conscience, 184
consciousness, 2–4, 79, 109. *See also* mem-
    ory; mind
Cook, M., 197
counting, cognitive basis, 114
Cro-Magnon people, 15, 111
Cronin, John E., 185
Crook, John, 36
cuckoldry, 11
cuisine, 11–12
Culicover, Peter W., 205
Culliton, Barbara J., 192
cultural determinism, 174
culture: definition, 77, 120, 187–188; di-
    versity, 124–125, 135–138, 151; relation
    to heredity, 19–21, 56–57, 72; relation
    to mind, 130–133, 148; scientific study,
    56–57, 123–124; transmission, 57–62
culturgen, 121

Daly, Martin, 190
dance, 101
Dani, New Guinea, 67
Dart, Raymond, 8
Darwin, Charles, 160–162, 209
Darwinism, 24, 30. *See also* natural selec-
    tion
Davis, Bernard D., 192
Davis, P. J., 188
Dawkins, Richard, 36
DeFries, J. C., 198
design, preference, 67–68

Desor, J. A., 197
DeVore, Irven, 195
dinosauroids, 155, 157
dinosaurs, 155–156
division of labor, 11
DNA, 55, 57, 194–195
Dodd, S. C., 207
Draper, Patricia, 195
dualism, 75
Dubois, Eugene, 8
Durham, William, 48, 206
Durkheim, Émile, 47

Eccles, John, 75, 198
Edelson, Edward, 198, 210
Edey, Maitland, 187
Eddington, Arthur, 63
Ehrman, Lee, 192
Eibl-Eibesfeldt, Irenäus, 36, 69–70, 187,
    197
eidylons, 53–54, 114
Ekman, Paul, 68–69, 197–198
Engels, Friedrich, 120, 206
epigenetic rules: defined, 70–72; relation
    to ethics, 173–184; relation to gene-cul-
    ture coevolution, 117, 184; relation to
    language, 112; relation to memory,
    82–83
Eskimos, 10
d'Espagnat, B., 193
ethics, 173–184
ethnographic curve, 124, 134–138,
    144–146
ethnography, *see* culture
ethology, defined, 23–24
Ettlinger, G., 204
eugenics, 39
evil, 173. *See also* ethics
evolution: four main steps, 7; natural se-
    lection, 24–26. *See also* gene-culture
    coevolution
evolutionary theory, 2, 24–36
eyebrow signal, 121

facial expression, 68–69, 115
Fagan, J. F. III, 196
Fagen, Robert M., 200, 202
Falk, Dean, 186, 209
Fantz, R. L., 197
Farb, Peter, 187